Designing Mathematics or Science Curriculum Programs

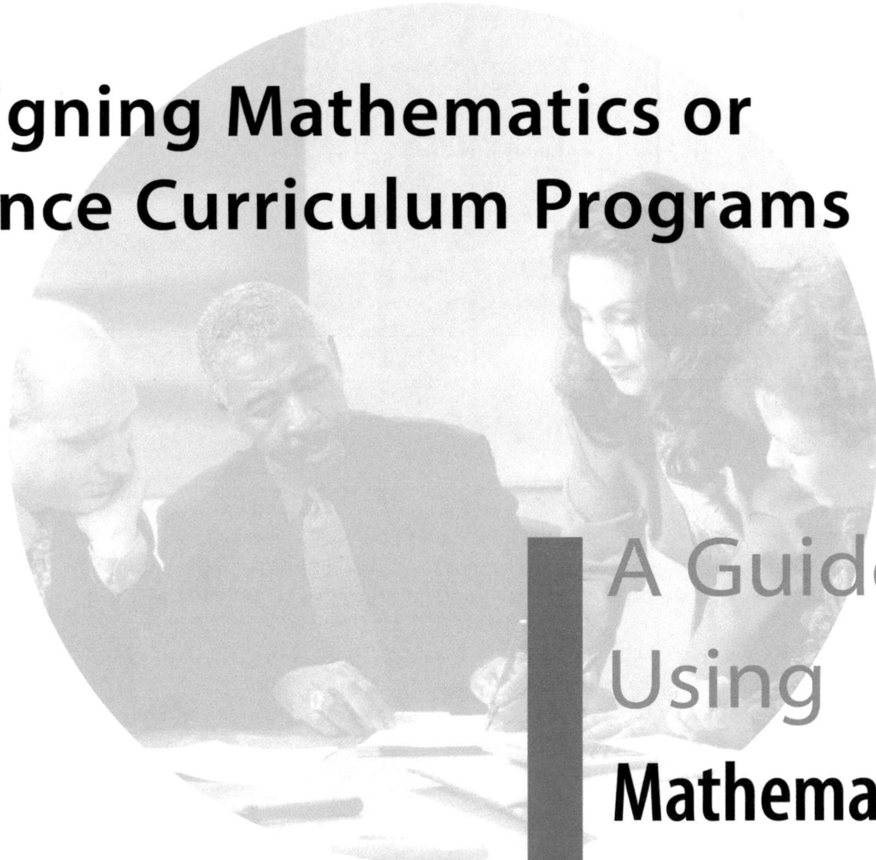

A Guide for Using Mathematics and Science Education Standards

Committee on Science Education K-12
and the Mathematical Sciences
Education Board

Center for Science, Mathematics,
and Engineering Education
National Research Council

National Academy Press
Washington, D.C.

NOTICE: The project that is the subject of this report was approved by the Governing Board of the National Research Council (NRC), whose members are drawn from the councils of the National Academy of Sciences, the National Academy of Engineering, and the Institute of Medicine. The members of the committee responsible for the report were chosen for their special competences and with regard for appropriate balance.

The NRC is the operating arm of the National Academies, which includes the National Academy of Sciences, the National Academy of Engineering, and the Institute of Medicine. The NRC was organized in 1916 by the National Academy of Sciences to associate the broad community of science and technology with the Academy's purposes of furthering knowledge and providing impartial advice to the federal government. Functioning in accordance with general policies determined by the Academy, the council has become the principal operating agency of both the National Academy of Sciences and the National Academy of Engineering in providing services to the government, the public, and the scientific and engineering communities. The Council is administered jointly by both Academies and the Institute of Medicine. Dr. Bruce Alberts, president of the National Academy of Sciences, and Dr. William A. Wulf, president of the National Academy of Engineering, are chairman and vice chairman, respectively, of the NRC.

The Center for Science, Mathematics, and Engineering Education (CSMEE) was established in 1995 to provide coordination of all the NRC's education activities and reform efforts for students at all levels, specifically those in kindergarten through 12th grade, undergraduate institutions, school-to-work programs, and continuing education. The Center reports directly to the Governing Board of the NRC.

This study by the Mathematical Sciences Education Board and the Committee on Science Education K-12 of CSMEE was conducted under a grant from the National Science Foundation (ESI-9355774) to the National Academy of Sciences/NRC. Any opinions, findings, and conclusions or recommendations expressed in this report are those of the authors and do not necessarily reflect the views of the National Science Foundation.

Designing Mathematics or Science Curriculum Programs: A Guide for Using Mathematics and Science Education Standards is available for sale from the National Academy Press, 2101 Constitution Avenue, N.W., Lock Box 285, Washington, DC 20055. Call (800) 624-6242 or (202) 334-3313 (in the Washington, DC metropolitan area). It is also available at http://www.nap.edu.

International Standard Book Number 0-309-06527-5

Printed in the United States of America.

THE CENTER FOR SCIENCE, MATHEMATICS, AND ENGINEERING EDUCATION COMMITTEE ON SCIENCE EDUCATION K-12

MATHEMATICAL SCIENCES EDUCATION BOARD

COMMITTEE TO DEVELOP A GUIDE FOR DESIGNING MATHEMATICS AND SCIENCE CURRICULUM PROGRAMS

Cathy Seeley (*Chair*), Texas Statewide Systemic Initiative, University of Texas at Austin-Dana Center, Austin, TX
Judy L. Capra, Jefferson County Public Schools, Wheat Ridge, CO
Glenda T. Lappan, Connected Mathematics Project and Department of Mathematics, Michigan State University, East Lansing, MI
Jack Price, Center for Mathematics and Science Education and Equity and Technology, California State Polytechnic University, Pomona, CA
Scott Samsel Stowell, Spokane School District No. 81, Spokane, WA

NATIONAL RESEARCH COUNCIL STAFF

Rodger Bybee, Executive Director, Center for Science, Mathematics, and Engineering Education (CSMEE)
Joan Ferrini-Mundy, Associate Executive Director, CSMEE; Director, Mathematical Sciences Education Board (MSEB), CSMEE
Susan Loucks-Horsley, Division Director, K-12 Policy and Practice, CSMEE
Harold Pratt, Director, Science Projects, CSMEE
Tom Rowan, Project Consultant
Kathleen Johnston, Project Consultant
Alfred G. Young, Administrative Assistant, K-12 Policy and Practice, CSMEE
Terry K. Holmer, Senior Project Assistant, Undergraduate Science Education, CSMEE
Sally S. Stanfield, Editor, National Academy Press
Dawn M. Eichenlaub, Production Manager, National Academy Press

REVIEWERS

This report has been reviewed in draft form by individuals chosen for their diverse perspectives and technical expertise, in accordance with procedures approved by the National Research Council's (NRC's) Report Review Committee. The purpose of this independent review is to provide candid and critical comments that will assist the institution in making the published report as sound as possible and to ensure that the report meets institutional standards for objectivity, evidence, and responsiveness to the study charge. The review comments and draft manuscript remain confidential to protect the integrity of the deliberative process. We wish to thank the following individuals for their participation in the review of this report:

Rick Billstein, University of Montana, Billings, MT
Jere Confrey, University of Texas at Austin, Austin, TX
Walter Denham, Mathematics Education Consultant, Fair Oaks, CA
Alice Krueger, High Plains Consortium, Aurora, CO
Michael Lang, Phoenix Urban Systemic Initiative, Tempe, AZ
Lyn Richbart, New York State Education Department, Albany, NY
Cary Sneider, Boston Museum of Science, Boston, MA
Bonnie Ward, Chestertown, MD, Public Schools, Chestertown, MD

While the individuals listed above have provided many constructive comments and suggestions, responsibility for the final content of this report rests solely with the authoring committee and the National Research Council.

Contents

Preface

Because the curriculum a student studies has a major effect on what she or he will learn in school, the design and implementation of curriculum programs is a high priority for local, state, and federal education agencies. Accordingly, the National Science Foundation (NSF) has designated the use of high-quality, standards-based mathematics and science curricula as important components of the systemic initiatives it funds at the local, state, and regional levels. The NSF not only wants states and districts to select high-quality instructional mathematics and science materials but also to assemble them in a coherent way within and across grade levels in a manner that has the greatest potential for improving student achievement.

The NSF's interest in enhancing student learning in mathematics and science motivated the agency to ask the National Research Council's (NRC) Center for Science, Mathematics, and Engineering Education (CSMEE) to develop a guide to help state- and district-level mathematics and science curricula developers design multi-year curriculum programs. In making its request, the NSF was aware of the importance and difficulty of developing coherent curriculum programs that provide students with the opportunity to learn mathematics and science in a continuous, interconnected, and cumulative way throughout their schooling, kindergarten through high school. The NSF asked for the guide to be based on the National Council of Teachers of Mathematics' (NCTM) *Curriculum and Evaluation Standards for School Mathematics*[1] and the NRC's *National Science Education Standards* (NCTM, 1989; NRC, 1996b). The guide was to be designed for use with local and state frameworks and curriculum guidelines by school districts engaged in the work of standards-based improvement efforts. Generally, districts have a number of instructional materials from which to choose but little or no guidance in

[1] In 1998, the NCTM released a discussion draft of its latest standards effort. It is entitled *Principles and Standards for School Mathematics: Discussion Draft* (NCTM, 1998).

determining how to assemble the materials to create coherent curriculum programs in either mathematics or science.

After CSMEE's receipt of the NSF's request, the Third International Mathematics and Science Study (TIMSS) began to release the results of its assessments of and contextual research into three student populations at the fourth- grade, eighth-grade, and the final year of secondary school levels. In addition to student achievement, TIMSS specifically collected data on the curriculum documents and textbooks used in the various nations participating in the study. Data were also collected from the results of surveys of students, teachers, school officials, and national officials. The mathematics and science curricula in the countries involved in the study were thoroughly analyzed, and a number of inferences were made about the relationships between student achievement and the curriculum. Although a causal connection has not been established between the TIMSS achievement results and curriculum, the possible relationship raises provocative questions about the mathematics and science curricula in the United States and provides new lenses through which to study and, hopefully, to improve them.

In response to the NSF's request for a curriculum program design guide, the Mathematical Sciences Education Board (MSEB) and the Committee on Science Education K-12 (COSE K-12), both within CSMEE, joined forces to form the Committee to Develop a Guide for Designing Mathematics and Science Curriculum Programs. The members of this committee are listed on page v of this report.

In developing this guide, the committee found a common thread between the NSF's concern that curriculum programs lack coherence and questions raised by TIMSS data about the mathematics and science curricula in the United States. In responding to this finding, this guide takes the position that learning is cumulative over time and, therefore, curriculum programs in the United States should be designed to be more coherent over time. This guide outlines the components of coherent programs in mathematics or science, proposes criteria by which their quality can be judged, and suggests a process for developing them.

Cathy Seeley
Chair, Committee to Develop a Guide for Designing Mathematics and Science Curriculum Programs

Harold Pratt
Study Director
Center for Science, Mathematics, and Engineering Education

Designing Mathematics or Science Curriculum Programs

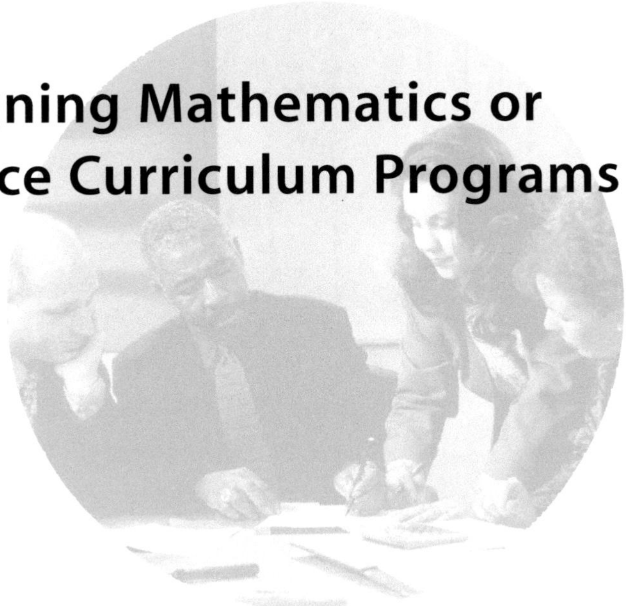

Introduction

PURPOSE OF THIS REPORT

The purpose of this report is to help those responsible for curriculum decisions in school districts to design coherent mathematics and science curriculum programs. Coherence, as it is used in this report, refers to the connectedness and sound development of ideas and skills presented to students within a year and over several years. To achieve coherence, a curriculum program must build new ideas and skills on earlier ones within lessons, from lesson to lesson, from unit to unit, and from year to year, while avoiding excessive repetition. As students construct and develop new ideas and skills, the concepts and processes they learn become richer and more complex.

In this report, a curriculum program[2] is defined as the content of instruction and the ways in which it is structured, organized, balanced, and delivered in the classroom over an extended period of time (at least a few years). The components of a curriculum program addressed by or discussed in this report are goals, standards, a common vision, a curriculum framework, and instructional materials.[3] (Teaching strategies and the classroom assessments associated with the use of instructional materials are also part of the program but will not be discussed in this report per se.)

All of the components of a curriculum program together are often referred to as the "intended curriculum" of a school district and may differ from what is implemented by the teacher—the "implemented curriculum"—and what is learned by the students—the "achieved curriculum." (This report does not deal with implemented or achieved curricula but monitoring both can provide districts with important data to guide their process and results.)

Although curriculum programs, frameworks, or guidelines of some nature are in place in states and school

[2] Use of the word "program" is meant to emphasize the multi-year nature of the curriculum and to differentiate it from the broadly used word, "curriculum."

[3] Instructional materials are the discrete physical components or blocks of curriculum including, for example, textbooks, software, kits, and teacher's guides.

districts across the country, they typically do not span all the years of precollege education and can lack coherence. Rather, local school districts as well as entities that can influence them, such as textbook companies, national curriculum development groups, and state educational agencies, tend to organize instruction around smaller "chunks" of schooling that may focus on the primary years (grades K-2), on grades 6-8, or on a particular high-school-level course. These smaller chunks of schooling are not necessarily designed to maximize the opportunities for all students to learn the content called for in standards. By contrast, this report emphasizes the importance of defining and coordinating curricula across the entire 13-year span—based on standards in use by local school districts—as a way to improve the quality of education. It describes the components of coherent curriculum programs based on mathematics or science standards or both and a process for designing such programs. [4] Through the process, schools and school districts will be able to develop greater alignment between existing curriculum programs and content standards. In many districts, curriculum program design committees will be formed to perform the task.

CURRICULUM PROGRAMS: A NATIONAL PERSPECTIVE

TIMSS and Curriculum Programs in the U.S. The Third International Mathematics and Science Study (TIMSS) compared the achievement of over 500,000 fourth-, eighth-, and twelfth-grade students in 41 countries. In addition to measuring achievement in mathematics and science, TIMSS measured the opportunities of students to learn mathematics and science. The data were obtained through classroom observations, videotaping of classroom teaching, teacher and student surveys, and an extensive review of each country's curricula. The results of the study indicated that the typical mathematics and science curricula in United States school systems are not well designed.

When fourth-grade student achievement data from TIMSS were analyzed, only one other country's students outperformed U.S. students in science. U.S. students also were above the international average in mathematics. However, when eighth-grade student

---

[4] The design of interdisciplinary curriculum programs for mathematics and science is not addressed in this report per se; however, the report's guidelines could be used for this purpose.

achievement data were analyzed, U.S. students scored below the international mean in mathematics and just above the mean in science. U.S. students slipped even farther at the twelfth grade, where, in science and mathematics general achievement, they outperformed only one other country's students.

There may be a number of reasons that the science assessment performance of the fourth-grade students in the United States was relatively high compared to those of students in other countries, including the fact many other countries do not include formal science instruction in the early grades; however, education analysts found it most startling that achievement appears to decline over time in the United States relative to other countries. Whatever the cause, it appears to be cumulative, contributing to the decrease in U.S. scores as schooling progresses through the grade levels.

One possible cause for this decrease is the nature of the curricula that many U.S. students experience over their 13 years in the schools (Schmidt et al., 1998). Most mathematics and science curricula in the United States lack coherence and focus, and that has caused some researchers associated with TIMSS to characterize the typical curriculum in the United States as a "mile wide and an inch deep" (Schmidt et al., 1997). When Schmidt compared the number of topics in U.S. textbooks and curriculum guides with those of other countries, he found that textbooks in the United States contained considerably more. As an example, Figure 1 displays data obtained in the analysis of science textbooks for three age levels. In addition, in the United States, fourth-grade mathematics and science textbooks contain an average of 530 and 397 pages, respectively, whereas in Japan, mathematics and science text-

Figure 1. Number of Science Textbook Topics for Three Age Levels Studied in TIMSS

Country	Number of Topics		
	9-year-olds	13-year-olds	High-school completion
United States	56	67	53
Japan	11	8	17
International Mean	25	27	23

books intended for the same age level contain an average of 170 and 125 pages, respectively (Schmidt & Valverde, 1998).

The breadth of topic coverage and lack of focus in the textbooks in the United States as illustrated by these data do not allow students to develop deep understanding of the topics. An analysis of eighth-grade physical science textbooks on the topic of chemical changes illustrates the problem (see Figure 2) (Schmidt et al., 1997). In the textbooks

Figure 2. Grade 8 Science Textbook Performance Expectations for "Chemical Changes"

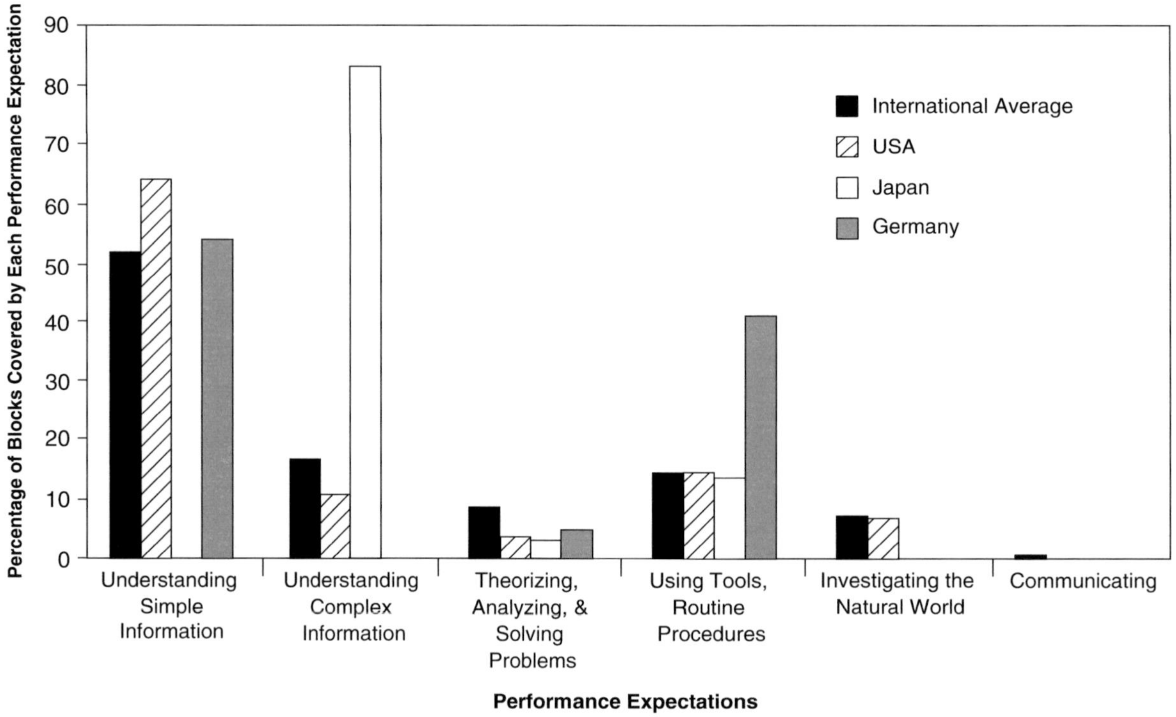

Reprinted from *A Splintered Vision: An Investigation of U.S. Science and Mathematics Education*, pg. 104, Exhibit 40, by permission of Kluwer Academic Publishers (Schmidt et al., 1997).[5]

[5] Percentages in some categories were so low or negligible that they may not appear on Fig. 2. Performance expectations as defined in the TIMSS study are as follows: **Simple information** (information such as vocabulary, facts, equations, simple concepts; examples include defining, describing, naming, quoting, reciting, etc.; specific examples are defining scientific terms [boiling point, niche], knowing symbols [abbreviations for units, chemical symbols], describing simple concepts [materials expand when heated, characteristics of animals]); **Complex information** (information involving the integration of bits of simple information; examples include differentiating, comparing, contrasting, synthesizing; specific examples are understanding how increased external pressure raises boiling point of liquids, how fire is a part of the life cycle of pine trees) (Robitaille et al., 1993).

used in the United States, approximately 65% of the textbook content calls for students to understand only simple information and only 10 % of the text-book content requires students to understand complex information. In the books used in Japan, 83% of the text-book content requires students to understand complex information, and there is virtually no emphasis on under-standing only simple ideas. Clearly, the performance expectations of textbooks in the United States as shown here are quite different from those in books used in Japan.

These data on the number of topics and level of expectations are an indica-tion of the lack of focus and coherence in curriculum programs in the United States. As Schmidt and Valverde (1998) state,

> "Attempting to cover a large number of topics results in textbooks and teaching that are episodic. U.S. textbooks and teachers present items one after the other from a laundry list of topics prescribed by state and local district guides, in a frenzied attempt to cover them all before the school year runs out. This is done with little or no regard for establishing the relationship between topics or themes on the list. The loss of these relation-ships between ideas encourages children to regard these disciplines as no more than disjointed notions that they are unable to conceive of as belonging to a disciplinary whole."

These findings from TIMSS serve to emphasize what many mathematics and science educators have long believed—that the typical curricula in schools in the United States are shallow yet over-loaded, undemanding, fragmented, unfocused, and incoherent. The TIMSS achievement data suggest that the toll this takes on student learning is great—and greater as the curricula advance through the grades.

Practices that Contribute to the Lack of Coherence in Curriculum Programs. A number of practices in curriculum design and implementation appear to contribute to the lack of coherence—and challenge—in math-ematics and science curricula in the United States.

- **Mastery.** In psychology, the concept of "mastery" derived from the era of behavioral objectives. Its use in mathematics education refers to a stimulus-and-response approach to instruction, with reinforcement of right answers, usually under condi-tions of speed. When and if such an approach produced high degrees of accuracy, the student was said to have mastered the concept. Delayed testing often revealed a significant loss in mastery. In addition, behavior-ist approaches did not support under-lying learning theory nor did they examine systematically the patterns of errors that can explain student difficulty. This led most mathematics educators to the belief that a direct and sole emphasis on mastery—as learned in isolation and established at a single moment—is a questionable practice. Too often, mastery, as defined previously, remains a major goal in many classrooms. As a result, when a student is not able to recall immediately a concept or procedure, often in a situation free of any context such as a drill-and-practice exercise, this is interpreted as a lack of "mastery." One consequence is that the same procedures and content are re-taught each year, often with minimal improvement in student outcomes. Another consequence is that, when mastery is a major goal yet students fail to achieve it, new concepts and procedures are delayed or not taught at all. Instead, as students are exposed to an annual cycle of repeating what was previously taught, they lose motivation as well as are denied access to higher level concepts, procedures, and problems. Students who are slower to gain skills early are especially hard hit by this practice because the impact of denied access to new concepts begins so early and accumulates over time, causing these students to fall farther and farther behind.

Rejection of direct and sole emphasis on mastery, where skills are learned in isolation and dependent on repetition, in conditions of timed tests, should not be interpreted as rejection of the important characteristics of procedural fluency and automaticity. Because many more complex skills and concepts in mathematics and science depend on simpler ones, a student is hampered if ease and quickness are not achieved. This can be accomplished through repeated practice, especially as embedded in other activities, and through careful assessment and accountability practices.

- **The practice of rote memorization.** It is difficult to change the attempt to achieve mastery through repetition and rote memorization because such practices are widespread and deeply engrained. Many teachers experience pre-college mathematics and science instruction that is based largely on rote memorization of disconnected facts and skills. Many never go on to acquire additional education or experience in mathematics and science strong enough to give them the confidence to teach these subjects in an appropriate manner. It is easier for teachers who have experienced traditional instruction and who have weak backgrounds to teach low-level skills over and over than it is for them to provide instruction that develops new concepts and procedures. Unfortunately, it is not easier for students to learn well this way.

- **Textbook content coverage.** Publishers of mathematics and science instructional materials at all levels often attempt to cover all possible content in single products—primarily textbooks—in order to meet the requirements of as many districts and states as possible. The result is a smattering of all possible concepts and skills so that each state or district can find the topics required by its syllabus or standards (Tyson, 1997). In addition, at the secondary level, long-standing practice and tradition have dictated much of the textbook content for decades. As a consequence, the secondary science course content recommended by the "Committee of Ten" in 1892 still prevails today virtually intact (Hoffman & Stage, 1993).[6] These courses mimic the design of freshman college courses in name and often in content structure and organization.
- **Overly flexible design.** A major curriculum design assumption in many local districts and states is that the curriculum should be flexible enough to allow for a choice of instructional materials by schools and teachers. In both mathematics and science, this has led to the use of the "module"—usually six to nine weeks in length—as the basic unit or building block of a curriculum. Several major curriculum development projects have produced complete K-6 programs based on a matrix that assigns four modules per grade level. Even when these programs are based on a well-thought-out development of concepts, processes, and skills over the six or seven elementary grades, schools, school districts, and states often "mix and match" the modules as they see fit. In this case, there should be a well-designed, multi-year framework within which the selected modules fit and can be coherently linked; otherwise, there is the danger that their inherent coherence and developmental sequence will be lost. Similar types of modular programs are emerging for use at the middle-grade levels from some of the same curriculum development groups. Districts may be tempted to mix and match these components, as well. The same caution about the implications of such a temptation for coherence and developmental sequence applies.

---

[6] The "Committee of Ten" was a committee of university presidents that met to consider the science preparation needed for college admissions.

- **Lack of attention by schools to coherent curriculum program development.** Most school districts do not devote sufficient economic resources and expertise to the development of curriculum programs. Typically, school districts use standards (or some other criteria) as the basis for the selection of textbooks and other instructional materials, and then these materials become the curriculum program for the district. As mentioned above, in some cases, the instructional materials may have been designed to provide coherence from unit to unit and from grade to grade. However, it is often the case that units, grade-level materials, and textbooks from different sources—even when they are internally well designed—constitute separate collections that do not relate well to one another. Also, the instructional materials selection process often does not include attention to multi-grade coherence and articulation. The curriculum program should guide the selection of instructional materials to ensure that the entire set provides the coherence needed to ensure the development of concepts and skills within a given grade level, from one grade level to the next, from one year to the next, and from elementary school to middle school to high school.

CURRICULUM PROGRAMS: THE POTENTIAL

With the development of national standards for mathematics and science and of standards-based curricula, educators have increased their understanding of practices that will contribute to coherence in curriculum programs and, of critical importance, enhance accessibility for all students. The potential for putting that understanding to work and thereby significantly improving the efficacy of curriculum programs for all students is discussed below.

National Standards Address the Importance of High-Quality Programs. Of the many landmark national standards documents that describe what students should know and be able to do by the end of their K-12 education, the first was the *Curriculum and Evaluation Standards for School Mathematics* (NCTM, 1989). It was followed a few years later by the *Benchmarks for Science Literacy* (*Benchmarks*) (American Association for the Advancement of Science [AAAS], 1993) and the *National Science Education Standards* (*NSES*) (NRC, 1996b). NCTM plans to release a new version of its standards, *Principles and Standards for School Mathematics,* in the year 2000. All versions of these standards make the case for improving and aligning all parts of the education system. See Table 1 on pg. 10.

Since the release of these documents, state educational agencies and local school districts have been developing their own content standards as a foundation of their efforts to improve the quality of educational opportunities for their students in K-12 mathematics and science (Council of Chief State School Officers [CCSSO], 1997). Standards related to curriculum programs generally have not been included in these content standards, although suggestions concerning curricula are often included in other documents, such as state frameworks.

This report assumes that schools and districts will be guided by the state and/or local district standards that they are required to follow and that they will also seek additional guidance from other state documents and national standards from the NCTM, NRC, and AAAS.

Coherence and Accessibility in Curriculum Programs. In describing high-quality mathematics and science programs over the 13-year period of K-12 schooling, national standards stress the importance of good design, articulation, and coherence in the K-12 curriculum program. Indeed, this emphasis has received particular focus in the revised NCTM *Standards* entitled *Principles and Standards for School Mathematics: Discussion Draft,* which include a mathematics curriculum principle: "Mathematics instructional programs should emphasize important and meaningful mathematics through curricul[a] that are coherent and comprehensive" (NCTM, 1998).

The connectness and sound development of ideas and skills over the years of schooling in a coherent program is often compared to the progression of a good story. Students become aware of and understand the connections between ideas as the story develops over days, months, and years. Of course, in mathematics and science, as in most core curriculum subjects, not just one story is involved. In its best form, a coherent program is one in which there are many stories: some are being told simultaneously, and there is an interdependence between them; some are unfolding progressively, with the complexity and level of conceptual understanding increasing from one segment or story to the next.

To achieve coherence, a curriculum program must (1) focus on the important ideas and skills that are critical to the understanding of important phenomena and relationships and that can be developed over several age levels; (2) help students develop an understanding of these ideas and skills over several years in ways that are logical

Table 1

The *National Science Education Standards* (NRC, 1996b) contain "Program Standards" that describe the conditions needed for high-quality school science. These standards include

- consistency across all elements of the science program and across the K-12 continuum;
- quality in the program of studies;
- coordination with mathematics;
- quality resources;
- equitable opportunities for achievement; and
- collaboration within the school community to support a quality program.

Specifically, the narrative for the second standard calls for "district-wide goals and expectations for student achievement, as well as the curriculum frameworks, [which] serve to ensure coherence and articulation across grades...".

The NCTM *Standards* (NCTM, 1989) promote essentially the same points in Standards 11, 12, and 13 of the "Evaluation" section, as follows:

Standard 11— Indicators of a mathematics program's consistency with the *Standards* should include

- student outcomes;
- program expectations and support;
- equity for all students; and
- curriculum review and change.

Standard 12— In an evaluation of a mathematics program's consistency with the *Standards*, the examination of curriculum and instructional resources should focus on

- goals, objectives, and instructional methodology;
- relative emphases of various topics and processes and their relationships;
- instructional approaches and activities;
- articulation across grades;
- assessment methods and instruments; and
- availability of technological tools and support materials.

Standard 13— In an evaluation of a mathematics program's consistency with the *Standards*, instruction and the environment in which it takes place should be examined, with special attention to

- mathematical content and treatment;
- relative emphases assigned to various topics and processes and the relationships among them;
- opportunities to learn;
- instructional resources and classroom climate;
- assessment methods and instruments; and
- the articulation of instruction across grades.

and that reflect intellectual readiness; (3) explicitly establish the connections among the ideas and skills in ways that allow students to understand both ideas and the connections among them; and (4) assess and diagnose what students understand to determine the next steps in instruction.

A coherent program does not contain simply a listing of vocabulary or content topics. Rather, ideas and skills connect and build on one another, i.e., they are clearly described; they include some indication of the level of performance expected of the students; and they are connected in a logical progression of ideas and skills. It is important that the standards and benchmarks used in the program contain enough detail to make clear the connections between lessons, units, or levels, and that the connections make increasingly rigorous development of ideas possible. (The second section of this report includes an expanded discussion of these important characteristics of standards and benchmarks.)

Connections can best be made among ideas and skills that are well understood and not just memorized as definitions or procedures that are quickly forgotten. Therefore, while a coherent curriculum program will usually contain fewer topics than an incoherent program, the topics will be richer and lead to greater depth and persistence of understanding.

If a deep understanding is to be achieved, content must be presented to students at an age when they have a readiness for it, are capable of understanding it, and can see the relationships between the ideas to which they are being exposed presently and those to which they were exposed previously (Schoenfeld, Smith & Arcavi, 1993).

Because ideas build and connect over time in a coherent program, monitoring student progress through thoughtful classroom assessment is essential to prevent the thread of ideas or skill development from being broken. Assessment used by teachers as feedback to guide the modification of the teaching and learning activities (often called formative assessment) has been demonstrated as an effective means to improve the achievement of students (Black & Wiliam, 1998).

It may seem from the discussion so far that, if a program has coherence, then only students who have successfully experienced everything preceding a unit can learn the intended outcomes of that unit. Such a view has the potential to limit access for many students. A belief that a student can move on to the next skill in a continuum only when that student has "mastered" all previous skills has prevented many students from ever experiencing the level or type of content in which they would be motivated to succeed. This has been especially true in mathematics, where the curriculum traditionally has been viewed as a linear progression of skills and procedures (NCTM, 1989).

It is important to stress that a coherent program should be accessible to all students. While the curriculum should be designed so that each learning activity builds on previous activities, instruction should be guided by decisions that allow every student, regardless of past experience, to participate in intellectually stimulating ways and to demonstrate continual progress. If the curriculum has been designed with rich, engaging tasks, appropriate instructional decisions can be made to assist all students in attaining significant cognitive growth. Use of a variety of tools, such as calculators, computers, measurement equipment, and computational recording devices, can increase the likelihood that students with past gaps in experience, particularly computational gaps, will not be denied access to new material while being encouraged to gain procedural fluency with past material. Students at all levels of preparation solve problems in their own unique ways and can contribute significant insights to a class's mathematical activity. Frequently, it is when their contributions are solicited and recognized that these students recognize the value in gaining computational fluency and show marked gains in achievement. Simultaneously, other class members continue to progress academically.

In mathematics, in particular, when mastery procedures that can be performed mechanically, such as paper-and-pencil computation or factoring of binomials, dominate the curriculum, underprepared students have virtually been excluded from reaching classes that involve more interesting contextual problems. The assumption has been made that students must demonstrate proficiency in low-level skills before engaging interesting and challenging ideas and problem solving. In such a system, a student with gaps in low-level skills or computational proficiency is highly unlikely to succeed.

A well-developed, coherent curriculum program not only is designed to take advantage of important previous knowledge but to have multiple entry points to allow students who may have gaps in their previous knowledge to participate and learn rigorous content. At least one NSF-funded curriculum project is built on this premise, with units that evolve to increasing levels of rigor and sophistication with entry points for students with less than complete prior experience. All students have opportunities to be successful, including those who may not have experienced previous units (Lappan & Phillips, 1998).

In science, accessibility for all students is possible through the inquiry provided by laboratory investigations that allow all students to approach new concepts with a common set of concrete experiences. Subsequent analyses of experimental data and constructions of explanation will vary in sophistication from student to student depending on their previous learning. Although ideas and activities may build on previous activity, each new investigation presents new opportunities for students with gaps in their past experience to contribute to their team's or class's solution to the investigation. Students with less comprehensive prior preparation can still reach an acceptable level of understanding and success. Having access and success through the concrete experience of the investigations may enable a student to find renewed interest and achievement in the importance and application of previous concepts and skills that would have been impossible in a more didactic program.

ASSUMPTIONS UNDERLYING THIS REPORT

Four broadly accepted assumptions underlie this report:

1. Some knowledge is more fundamental than other knowledge.
2. Student learning can be significantly enhanced when learning experiences are designed in a coherent way based on what students have already learned.
3. Curriculum programs should specify what ALL students should know, understand, and be able to do.
4. The curriculum affects what is taught and learned.

Some Knowledge Is More Fundamental Than Other Knowledge. Some ideas and procedures are more fundamental than others because they are the foundation for ideas that will be taught, have rich explanatory power, and relate to everyday experiences (NRC, 1996b; NRC, 1993). Mathematics and science curriculum programs should focus on providing students with the opportunity to learn a limited number of fundamental ideas well rather than presenting them with a long list of random, unconnected information (Schoenfeld, Smith & Arcavi, 1993).

Student Learning Can Be Significantly Enhanced When Learning Experiences Are Designed in a Coherent Way Based on What Students Have Already Learned. Students are more likely to learn when the sequence of their experiences is designed so that the development of their understanding of ideas grows and expands over time, reaching higher and higher levels of sophistication and depth. Conversely, students' learning suffers when their experiences have no particular order and do not require or capitalize on earlier learning (NRC, 1999b). This does not mean that a student who has missed one or more units should not be allowed to progress. There are many alternative and creative ways to assist students to progress when some elements of important prior knowledge are missing.

Curriculum Programs Should Specify What ALL Students Should Know, Understand, and Be Able To Do. This assumption echoes the central theme of both the *NSES* and the NCTM *Standards*, i.e., the content described in both documents is for ALL students.[7] This report focuses on curriculum programs that will make that learning possible. Regardless of the source, an agreed-upon set of learning targets to be met by all students over a given amount of time (such as grades K through 12 or any other multi-year period) is a key element in the design of an effective curriculum program.

The Curriculum Affects What Is Taught and Learned. Teachers adapt and modify most curricula and instructional materials before and during use in the classroom. Even though the intended curriculum may vary from the delivered curriculum, there is ample evidence that the intended curriculum still has a significant effect on what is presented and learned (Schmidt et al., 1998).

In addition, the authoring committee acknowledges that the effort to improve mathematics and science education at the K-12 level must be embedded in a systemic context. In systemic reform, goals, standards, instructional materials, teaching practices, professional development opportunities, and assessment practices all are aligned with one another. In systemic reform, educational agencies adopt policies for the establishment and alignment of high-quality programs in curricula, teaching, assessment, professional development, and systems of support (Smith & O'Day, 1991; O'Day & Smith, 1993). The shaded and non-shaded portions of Figure 3 show key aspects of the system.

As noted in the "Introduction," this report addresses the steps leading to the design of the curriculum program, represented by shading in Figure 3. It does not address professional development, large-scale district, state, or national assessments, and support systems represented by the non-shaded

---

[7] The current version of the NCTM *Curriculum Standards* contains a few standards in grades 9-12 for advanced students (NCTM, 1989).

Figure 3. Using Standards to Improve Student Achievement

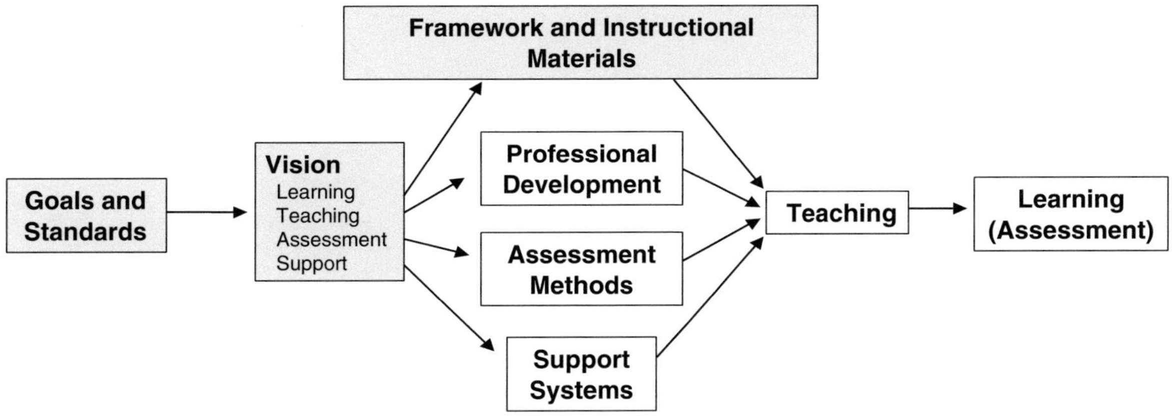

portions of Figure 3. These systemic components also play a role in curriculum programs but primarily in the implementation phase. While it is outside the scope of this report to provide a detailed examination of these components, their importance and their role in implementation are addressed briefly in an appendix (Appendix A).

Finally, an important premise of this report is that what a student learns depends to a great degree on how he or she is taught (NCTM, 1989). As school districts develop curriculum programs, they need to be as thorough in their consideration of and communication about the instructional approaches of teachers as they are in their consideration of and communication about mathematics and science content. Instructional approaches that will lead students to process information for meaningful interpretations and to think creatively about mathematics and science must be clearly expressed and be coherent from unit to unit and from grade to grade. While these are not topics of this report per se, instructional methodology must support student development of conceptual understanding or the intended levels of student achievement will not be attained.

The next section of this report, "Components of Coherent Mathematics and Science Education Curriculum Programs," outlines the components of a coherent curriculum program and describes the criteria necessary to design and create each component. It then suggests a process for using the components to develop the curriculum program.

Components of Coherent Mathematics and Science Education Curriculum Programs

A curriculum program should contain goals, content standards, a common vision, a curriculum framework, and instructional materials.[8] These components communicate the structure, organization, balance, and delivery of the content that students are expected to learn and, when effectively employed by teachers, that have the most direct impact on student learning in mathematics and science. These components constitute the instructional blueprint for the school district. This blueprint enables school system stakeholders—students, teachers, administrators, school board members, parents, representatives of colleges and universities, and others—to have a clear understanding of what students are expected to learn and the instructional opportunities they will have to learn it.

GOALS OF K-12 MATHEMATICS AND SCIENCE CURRICULUM PROGRAMS

Effective curriculum programs have goals that serve several important and interrelated functions. These goals should be written to communicate the overall purposes of the program to many audiences, including staff, parents, and policy makers, such as school board members. Goals also should be used to guide the actions and decisions of teachers, administrators, and support staff as these personnel develop, implement, and support activities to improve the quality of mathematics and science education in the district.

Goals serve to avoid the confusion between ends and means. For example, goals—and the standards that are

[8] The term "framework" refers to the "skeleton" or frame that guides and organizes the placement of the instructional materials. It does not refer to the comprehensive documents that many states have developed to describe all aspects of a mathematics or science program. Further, neither "assessment" nor "vision" is included as a component here. Assessments used by schools, districts, or states, although essential in the total scheme of improvement (see Fig. 3), are outside the definition of curriculum program and the scope of this report. Because high-quality instructional materials should contain a variety of assessments for use by classroom teachers, the presence of aligned assessments will be an important criterion in evaluating instructional materials. While a common "vision" is a critical starting point for the design of a coherent curriculum program, it is not often considered to be a part of the actual program blueprint.

derived from them—help to maintain a focus on significant ends or outcomes but do not dictate a single means as "the way" to approach instruction. A number of suitable and effective instructional strategies can be used to help students learn the mathematics and science content.

Goals for mathematics and science curriculum programs rarely stand alone. Often, they are derived from a broader set of educational goals already in place for the state or district. Groups designing curriculum programs may decide that their local goals should reflect the goals of the NCTM *Standards* (NCTM, 1989) and/or the *NSES* (NRC, 1996b).

The *NSES* begin with a "Call to Action." The first sentence of this call is an important and broad goal: "The nation has established as a goal that all students should achieve scientific literacy" (NRC, 1996b). That overarching goal clearly and unambiguously frames the other goals by stating that the standards are to achieve scientific literacy for ALL students. The document then sets out four learning goals that describe students who are able

- "to experience the richness and excitement of knowing about and understanding the natural world;
- to use appropriate scientific processes and principles in making personal decisions;
- to engage intelligently in public discourse and debate about matters of scientific and technological concern; and
- to increase their economic productivity through the use of the knowledge, understanding, and skills of the scientifically literate person in their careers."

The NCTM *Standards* state that "...today's society expects schools to insure that all students have an opportunity to become mathematically literate, are capable of extending their learning, have an equal opportunity to learn, and become informed citizens capable of understanding issues in a technological society" (NCTM, 1989). Following that opening charge, five goals for ALL students are given:

- that they learn to value mathematics;
- that they become competent and confident in their ability to do mathematics;
- that they become mathematical problem solvers;
- that they learn to communicate mathematically; and
- that they learn to reason mathematically."

Criteria for Goals

- *The target audience should be clearly specified.* The *NSES* and NCTM

Standards make it clear that their goals and standards are intended for all students. If the audience for a set of goals is not all students, this should be clearly stated. The absence of such a declaration creates uncertainty when subsequent decisions are made about standards, instructional materials, instructional grouping, assessment, budget allocations, and so on. Although the goals in the national standards are for all students, it should be recognized that different students will proceed at different rates and, therefore, will have some common and some individualized experiences. Any student willing to work hard and persistently should be able to have access to all aspects of the curriculum with no disadvantage due to race, ethnicity, religion, socioeconomic class, gender, or other nonrelevant feature.

- *Goals should communicate broad outcomes and values.* Usually goals convey the essence of intended student behavior, ability, and knowledge. Generally, goals are not measured directly but are transformed into progressively more specific and measurable outcomes that can be used to develop assessments. Several levels of specificity can be derived from the goals. Typically, the next higher level of specificity consists of content standards, often followed by performance standards that describe the level of expectations called for in content standards. Yet another level of specificity would be objectives that are more specific and unique to student activities; these are usually found in instructional materials.
- *Goals should provide guidance.* Although general in nature, goals should provide guidance for the decisions that shape the curriculum program. They should not be written in such broad terms or so ambiguously that they could be interpreted in almost any manner. For example, both the NCTM *Standards* and the *NSES* make the target of "all students" very clear. The mathematics standards call for students to become "mathematical problem solvers," an outcome that moves beyond the traditional goal of getting correct answers to arithmetic exercises. The *NSES* add personal decision making, engaging in public discourse, and economic productivity to the traditional "understanding the natural world" outcome. If used as guidance, goals such as these can significantly affect the nature of curriculum programs.
- *Goals should describe student outcomes, not system outcomes.* In this era of standards, emphasis on student learning, and use of student learning outcomes, goals should be written

with students as the focus. At present, goals often are written to describe outcomes for the district or school, such as expecting the scores on a standardized norm-referenced test to improve during the next year.

- *Goals should be widely accepted.* As with any policy document, goals will be of little value if they are not widely accepted. For maximum acceptance, goals should be free of jargon and highly technical language so that the public can understand them. To the degree possible, the development process should be initiated and led by the science and mathematics staff, but it should also involve extensive input, hearings, public and staff discussion, and review to achieve as high a level of agreement for the goals as possible.
- *Avoid frequent modification of goals.* After the goals are accepted, they guide the next steps in the process. The flow of this development process is in one direction. Once the total curriculum program has been implemented and is in place, modification of the goals should be undertaken with care, as unexamined changes could create inconsistencies throughout the program.

CONTENT STANDARDS

A comprehensive set of content standards that defines what students are to understand or be able to do is the key component in the design of an effective curriculum program. These content standards should be in place, along with goals, before development of the other components begins. The case for using standards to improve the quality of mathematics and science education and the achievement of students has been made by a number of groups (Gandal, 1997; Education Commission of the States [ECS], 1996; The Business Roundtable, 1996; McLaughlin & Shepard, 1995). The history of educational standards also has been well documented (Bybee, 1997; ECS, 1996).

Although the initial development of content standards in mathematics and science was conducted at the national level (NCTM, 1989; AAAS, 1993; NRC, 1996b), virtually every state has developed its own mathematics and science education standards modeled in various degrees after the national standards (CCSSO, 1997). In some local-control states, each district is responsible for the development of content standards, although model state standards often are available as a default alternative.

The *National Science Education Standards* define the content standards as "what students should know, understand and be able to do in natural science" (NRC, 1996b). The National Academy of Education has defined both content and performance standards, as follows: "Content standards are broad descriptions of the knowledge and skills students should acquire and be able to do in a particular subject area" and "Performance standards are the more specific, concrete examples and explicit definitions of what students must know and be able to do to demonstrate that they have attained the content standards. Performance standards should include multiple benchmarks that lead to higher standards of proficiency, making it possible to demonstrate progress for students at all levels of achievement. To the extent that content standards include examples of assessment tasks and of various levels of real student performance in response to these tasks, they can better provide teachers and the public with the insights into the concrete meaning of the standards and specific expectations for student learning" (McLaughlin & Shepard, 1995).

Although states and schools districts rarely develop both types of standards initially, understanding the purposes of both types and the distinctions between them is important before standards are written for curriculum programs. Content standards describe "what" is to be learned, and performance standards describe "how" it will be assessed. The mistake that many writers make is to attempt to make content standards serve both purposes by using words of action such as "describe," "explain," or "apply" to help describe content. Verbs such as these, if followed by detailed performance expectations, can be used as one means of assessing student knowledge and understanding, but they immediately eliminate other possible means of assessment and often obscure the important content or skill that should have been described in the content standard. The issue of the language used in the standards is visited again in the next section, "Criteria for Content Standards."

Criteria for Content Standards

Many groups faced with the task of designing curriculum programs will be expected to use a set of pre-existing content standards. Others may have the opportunity to create their own standards based on national or state standards. In this latter case, a few basic criteria or guidelines for writing standards will be useful.

- *A clearly specified audience.* The *NSES*, NCTM *Standards*, and most state standards make clear that they are intended for all students. Without a clear identification of audience, subsequent decisions about instructional materials, assessment, and budget allocations are made more difficult.

- *Fundamental content.* One of the important purposes of standards is to focus instruction and student learning on a deep understanding of fundamental content.

  In science, the *NSES* (NRC, 1996b) considers content to be fundamental if it

  - "represents a central event or phenomena in the natural world;
  - represents a central scientific idea and organizing principle;
  - has rich explanatory power;
  - guides fruitful investigations;
  - applies to situations and contexts common to everyday experiences;
  - can be linked to meaningful learning experiences; and
  - is developmentally appropriate for students at the grade level specified."

  In the national mathematics curriculum standards, content is considered fundamental if it meets the following criteria (NCTM, 1989):

  - "Knowing mathematics is doing mathematics. Informational knowledge has value in the extent to which it is useful in the course of some purposeful activity.
  - All students should have the opportunity to develop an understanding of mathematical models, structures, and simulations applicable to many disciplines.
  - Students should have a balanced approach to calculation, be able to use appropriate procedures (including mental arithmetic, paper-and-pencil procedures, and calculators or computers), find answers, and judge the validity of those answers.
  - The content of the curriculum should be appropriate for all students. All students should have an opportunity to learn the important ideas of the standards."

- *Scientific and mathematical accuracy.* All content standards should be reviewed carefully for scientific and mathematical accuracy.
- *Understandable vocabulary.* The general public should be able to read and understand the content standards. The terminology used should be similar to or reflect terminology used by the general public. The vocabulary for the lower grades should be adjusted to "signal the nature and sophistication of the understanding sought" (AAAS, 1993).
- *Appropriate grain size.* Many writers of standards are tempted to craft statements that contain several related ideas in an effort to present a coherent view

of the topic at hand. Other writers do the opposite and write several statements that parse ideas into their smallest parts. Both approaches have advantages and disadvantages. A collection of ideas may communicate how the components of the collection are related but also may require very large amounts (many weeks) of instructional time and the assessment of many different concepts or skills. On the other hand, confining one idea to one standard may make writing performance standards and assessments much easier, but one idea often cannot be evaluated alone; it must also be evaluated in light of student understanding of other ideas and their relationship or connections with one another.

- *Use of verbs to describe the outcomes.* [9] The language used to describe what students should know or be able to do should do just that. If the outcome is knowledge or understanding, the standard should contain the verbs "know" or "understand." If the outcome is a skill, language related to that skill (such as "measure," "communicate," or "design") should be used. The *NSES* developers settled on "understand" and "develop the abilities of" to communicate knowledge and skills, respectively, and the *Benchmarks* developers chose "know" and "know how." Both groups address the qualities of effective instruction in other ways: the *NSES* included "Science Teaching Standards," and Project 2061 included chapters on teaching in *Benchmarks* as well as in *Blueprints for Reform: Science, Mathematics, and Technology Education* (NRC, 1996b; AAAS, 1993 and 1998). In mathematics, the NCTM communicated its emphasis on instruction by developing a separate document, *Professional Standards for Teaching Mathematics* (NCTM, 1991c).

  Verbs that describe instruction, such as "investigate," "explore," or "research," should be avoided. These action verbs are often used to convey a message about the desirability of "hands-on" teaching, but, in so doing, they can limit instruction to a single strategy. Allowing standards to imply instruction also results in statements that communicate the "means" of learning rather than the "end" or outcomes of instruction.

  The writers of many state and local standards have used a variety of action verbs to describe different

---

[9] *Benchmarks for Science Literacy's* chapter on "Characterizing Knowledge" has an excellent discussion of this topic (pg. 312).

observable behaviors as well as levels of sophistication or understanding as a way to guide the assessment of student understanding of the content of the standard. The problem with this is that the choices of action and level are arbitrary and limit the performance that can be used in the assessment to the ones specified. A much more satisfactory solution is to write the content standards as statements of knowledge or understanding and skill and to develop a separate set of performance standards to guide the assessment process. (See the earlier discussion of performance standards on pg. 21.)

CURRICULUM FRAMEWORK

A framework is the listing of outcomes (usually content standards or benchmarks) by grade level that guides the development of the curriculum and the selection and placement of instructional materials. Ideally, the framework also includes the performance standards associated with the content standards. It is a key component of a coherent curriculum program. In some ways, a framework is like an elaboration of the scope-and-sequence documents that have traditionally served as a basis for many district programs. In this report, though, use of the term "framework" is meant to highlight the critical need for attention to coherence—to the way the content that students learn builds within a year and over a span of years.

Designing Frameworks to Facilitate Growth of Understanding. An important function of the framework is to identify and locate the standards or benchmarks that will facilitate the growth of understanding of ideas and skills, an important characteristic of coherence, during a year and from year to year. For example, in science, students should understand by the end of high school that, "The physical properties of [a] compound reflect the nature of the interactions among its molecules" (NRC, 1996b). Students could memorize this statement fairly quickly; however, they might understand the concept better if they first experiment with a variety of properties, know about the existence and properties of atoms, and understand that everything is made up of atoms. They also need to understand the relationship between atoms and molecules and realize that molecules vary in shape and size. *Designs for Science Literacy* describes such a sequence as follows: "...what is learned now should be based on what was learned earlier, what is capable of being learned now, and what needs to be learned next (AAAS, at press)." In short, in science, the more complex, higher level idea is understood best when it is built on a comprehension of earlier knowledge. Figure 4 on pg. 26 illustrates in more detailed fashion this growth of understanding from the *NSES*.

In mathematics, a similar observation about growth of understanding has been made by researchers involved in an in-depth NSF-funded study of how young children develop computational understanding based on their informal background knowledge (Carpenter et al., 1992).

The NCTM "Probability Standard" for grades 9-12 illustrates this. The goal within the standard is for students to "describe, in general terms, the normal curve and use its properties to answer questions about sets of data that are assumed to be normally distributed" (NCTM, 1989). This outcome builds on many experiences from earlier grades. These include constructing and interpreting a variety of graphs, including line graphs, and making connections among a data set, a graph, and numerical statistical representations, such as mean and standard deviation. To use the normal curve in the ways described, a student should understand what a normal distribution is and should have had experiences using data to test the validity of hypotheses. It also is best if students have learned about rational numbers and their applications in a variety of situations. Yet another expectation is understanding and skill in computation with rational numbers. Figure 5 on pg. 27 illustrates in a more detailed fashion the growth of understanding from the NCTM "Statistics Standard."

The treatment of the science and mathematics standards represented in Figures 4 and 5 represent only the "big idea" parts of these standards. It should be noted that the framework allows considerable flexibility for the design of the sequence of learning activities in the instructional materials. For example, in Figure 5, specific graphing forms, such as bar graphs, are not named at any of the levels. The grade levels at which such forms are introduced vary from one program or location to another. Some programs might wait until second grade and precede the introduction with student-invented graphing forms. The examples provided in Figures 4 and 5 do not indicate the precise placement of the standards. They are offered as an illustration of one thread through the curriculum program that can facilitate the growth in understanding of a fundamental concept. The national mathematics and science standards and the *Benchmarks* contain many such threads that are not obvious without developing the "growth of understanding" figures. These in turn facilitate the placement of the standards in the framework in a coherent manner.

Figure 4. Growth in Understanding the Structure of Matter: An Illustration of How Understanding Can Progress Over Many Years Based on the *National Science Education Standards*

By the end of grade 4, students will understand that

- Objects have many observable properties, including size, weight, shape, color, temperature, and the ability to react with other substances. Those properties can be measured using tools, such as rulers, balances, and thermometers.
- Objects are made of one or more materials, such as paper, wood, and metal. Objects can be described by the properties of the materials from which they are made, and those properties can be used to separate or sort a group of objects or materials.
- Materials can exist in different states— solid, liquid, and gas. Heating or cooling can change some common materials, such as water, from one state to another.

By the end of grade 8, students will understand that

- A substance has characteristic properties, such as density, a boiling point, and solubility, all of which are independent of the amount of the sample. A mixture of substances often can be separated into the original substances using one or more of the characteristic properties.
- Substances react chemically in characteristic ways with other substances to form new substances (compounds) with different characteristic properties. In chemical reactions, the total mass is conserved. Substances often are placed in categories or groups if they react in similar ways; metal is an example of such a group.
- Chemical elements do not break down during normal laboratory reactions involving such treatments as heating, exposure to electric current, or reaction with acids. There are more than 100 known elements that combine in a multitude of ways to produce compounds, which account for the living and non-living substances that we encounter.

By the end of grade 12, students will understand that

- Matter is made of minute particles called atoms, and atoms are composed of even smaller components. These components have measurable properties, such as mass and electrical charge. Each atom has a positively charged nucleus surrounded by negatively charged electrons. The electric force between the nucleus and electrons holds the atom together.
- Atoms interact with one another by transferring or sharing electrons that are farthest from the nucleus. These outer electrons govern the chemical properties of the element.
- An element is composed of a single type of atom. When elements are listed in order according to the number of protons (called the atomic number), repeating patterns of physical and chemical properties identify families of elements with similar properties. This "Periodic Table" is a consequence of the repeating pattern of outermost electrons and their permitted energies.
- Bonds between atoms are created when electrons are paired up by being transferred or shared. A substance composed of a single kind of atom is called an element. The atoms may be bonded together into molecules or crystalline solids. A compound is formed when two or more kinds of atoms bind together chemically.
- The physical properties of a compound reflect the nature of the interactions among its molecules. These interactions are determined by the structure of the molecule, including the constituent atoms and the distances and angles between them.

Figure 5. Growth in Understanding of Statistics: An Illustration of How Understanding Can Progress Over Many Years Based on the NCTM *Standards*

By the end of kindergarten, students will understand that

- Objects and data can be sorted and displayed so that population characteristics are seen and comparisons can be made between populations (e.g., Did more students wear shoes with laces, shoes with Velcro, or slip-on shoes? Was it a lot more or just a little more?).

By the end of grade 2, students will understand that

- Data are organized and displayed in a variety of ways to communicate information about groups or events (e.g., tables, charts, graphs, etc.).

By the end of grade 4, students will understand that

- Graphic displays of data are interpreted to describe the characteristics of things such as populations and events.
- The general shape of the data as seen in the display can often be used to get a general sense of the story the data are telling.
- Information from a graphic data display can be used to solve problems, to make interpretations, or to generate questions (e.g., Do you think that next month's weather graph will show the same patterns as this month's? How is this month's weather graph different from last month's? Why do you think that is so?).

By the end of grade 6, students will understand that

- Graphic displays of data can be used to make inferences and predictions and to support arguments about people and events.
- Looking at where the data cluster, the center, and how the data spread out can help with interpreting the story that the data tell.
- The way data are displayed has a significant effect on the ease of reading the display, as well as on the quality and accuracy of interpretations made by observers (e.g., If we group the data in intervals of 10, how is the display changed?).

By the end of grade 8, students will understand that

- Decisions about data collection and display should be based on the questions that are to be addressed by the data.
- Data displays can be constructed to communicate information accurately and clearly. They can also be designed to distort communication and encourage wrong interpretations.

By the end of grade 10, students will understand that

- Statistical methods provide powerful means of influencing people and making decisions based upon information.

By the end of grade 12, students will understand that

- Statistical methods are used to test a wide variety of hypotheses about populations and events.
- If appropriate, data can be transformed to facilitate their interpretation and the making of predictions based upon the data. Statistical measures of central tendency, variability and correlation are used to describe and make decisions about populations and events.
- Large populations can be sampled to gather data to describe the population and/or to make predictions. The sample size and sampling method affect the confidence that one has in the interpretations and predictions that are made about the population.

It also should be noted that, at the district and school levels, standards that span more than one grade level can be too broad to be useful. Because most schools are organized around one-year grade levels, standards may need to be specified further to fit one-year blocks. In some cases, even shorter increments may be needed. To accommodate these needs, the concepts in the content standards (state and/or local) that span several grade levels can be "unpacked" and assigned to particular grade levels. At times, adding intermediate standards or benchmarks will be desirable because the conceptual or procedural gap between standards is so great. This is particularly true at the earliest level (grade 4 for the NCTM *Standards* and the *NSES*; grade 2 for the *Benchmarks*). Because of the large grade span and even larger difference in cognitive growth, informal standards for kindergarten or the early primary grades may be useful, also. For similar reasons, it may be useful to find intermediate "standards" to fill intervals that may exist in state or national standards, especially at the higher-grade levels.

Although Figures 4 and 5 are linear, both contain a number of interconnected story lines. Anyone familiar with the concepts and processes of mathematics and science is aware of the multiple connections that exist among ideas within each discipline. The national mathematics and science standards describe some aspects of this type of connection, both within the text of the standards and in elaborations of the standards. Maps of the content standards across the grade levels, as illustrated by Appendixes B and C, can make even more apparent the comprehensive and interconnected nature of subject matter (Ahlgren & Kesidou, 1995). Such maps, whether they are simple or complex, represent the next possible phase in the development of a framework. This phase requires considerable time and effort. Project 2061 is planning to produce a series of maps based on the *Benchmarks* that illustrate such connections graphically (AAAS, 1999a).

Criteria for a Curriculum Framework

- *Concepts in the framework should be assigned to grade levels based on what students are capable of learning.*
- *The framework should indicate development of processes/skills/abilities over several years.*
- *The framework should clearly indicate what standards are prerequisites for other standards.*
- *The assignment of content across grade levels should be appropriately balanced or concentrated.* (For example, do those who have been tasked to design the curriculum program want to emphasize geometry at the tenth

grade or earth science at the eighth grade? Or is a coordinated or integrated selection of content desired at each level?)

- *Concepts in the framework should be grouped to form the basis of units or courses, with logical connections shown both within a grouping and across grades.*
- *The framework should account for all standards.*

In a well-designed curriculum program, the sequence is cumulative, with each subsequent level applying, extending, and building on previously obtained knowledge. The main purpose is to build coherence into the program by describing a reasonable flow of these ideas across the grade levels.

It should be clear that neither the standards nor the framework constitute a curriculum program. Standards and frameworks identify the concepts that are to be learned and the order in which they are to be addressed. They do not specify how the content is to be taught. This is addressed by the fourth—and final—component of the curriculum program to be addressed in this report—instructional materials.

INSTRUCTIONAL MATERIALS THAT SUPPORT THE K-12 STANDARDS

There is broad consensus in both the mathematics and science education communities that use of instructional materials aligned with the content and teaching standards in national standards documents is a critical component of effective curriculum programs. With funding from the NSF, developers have designed instructional materials to address the standards, have tested them in classrooms, and have made available evidence of their effectiveness. The use of exemplary instructional materials to support the student learning identified in the K-12 instructional framework increases the likelihood that all students will have an opportunity to attain the level of understanding called for in the standards.

In recent years, several groups have developed objective criteria that can be used to identify exemplary instructional materials. Various review instruments have been produced that can be used by a school or district to select high-quality instructional materials. While the review instruments vary in format and some criteria, they all address two important dimensions of materials: 1) the degree of alignment between the content of the materials with that specified in the standards; and 2) the quality of the suggested instructional strategies. Many of the instruments also include a review of the assessments that are used in the materials and their degree of alignment with the specified learning outcomes (AAAS, 1997; NCTM, 1995; NRC, 1999c;

National Science Resources Center (NSRC), 1998a; U.S. Department of Education [DoEd], 1997; NSF, 1997; National Association of Biology Teachers (NABT) [year of publication not available]; Kahle & Rogg, 1996).

Both the materials review process developed by Project 2061 (AAAS, 1997) and the process and instrument developed by the Center for Science, Mathematics, and Engineering Education at the NRC link the review of instructional strategies to specific learning outcomes (AAAS, 1997; NRC, 1999c). Although the two tools differ in many ways, both of them first examine the match between the materials under review and the learning outcomes of the *Benchmarks*, the *NSES*, or other relevant local standards. If materials do not match standards to a reasonable degree, they are dropped from further consideration for selection. If there is a reasonable match between the materials and content standards, the materials are examined more closely to judge whether or not the teaching strategies in the materials provide adequate opportunities for students to learn what is called for in the content standards.

An instructional materials review and selection process from the U. S. Department of Education differs from those discussed above in that it asks reviewers to ascertain empirical evidence that student learning can be attributed to use of the materials (DoEd, 1997).

Criteria for the Selection of Instructional Materials

The following broad criteria have been gleaned from the procedures for analyzing and selecting instructional materials described above.

- *The content of the instructional material should be mathematically and scientifically accurate, consistent with the outcomes in standards, and targeted at the level called for in the framework.*
- *The instructional strategies consistently used by the material should be supported by learning research and make it possible for students to attain the specific outcomes identified in the first criterion, above.*
- *The assessments that accompany the material should be aligned with the content in the standards and the level of understanding or skill expected by the standard.*
- *The support for the teacher in the teacher's guide and ancillary materials should be adequate.*
- *There should be evidence from field trials that students can learn the specified content and skills if the materials are used as intended.*

With these components and criteria in mind, a process for developing a complete curriculum program is discussed in the next section, "Process for Designing a Curriculum Program."

Process for Designing a Curriculum Program

The process of designing a curriculum program that includes components that meet the criteria described in the preceding section requires considerable time and commitment. Fortunately, the process does not have to be considered completely implemented for improvements in mathematics and science teaching and learning to be realized. Each stage of the process makes a contribution to these goals.

This report assumes that a curriculum program design committee, with representatives of various stakeholders in the school system or district, will be responsible for the design process. This process—described in this section—will be a major professional development experience for the committee members.

The process described in this section and illustrated by Figure 6 is not intended to be prescriptive but, rather, to suggest how to design a curriculum program. The boxes in Figure 6 represent key steps in the process. Factors that influence the process are represented by ovals.

Early in the process of developing or revising a curriculum program, the committee should study the mathematics and science education context of the

Figure 6. Process for Designing a Curriculum Program

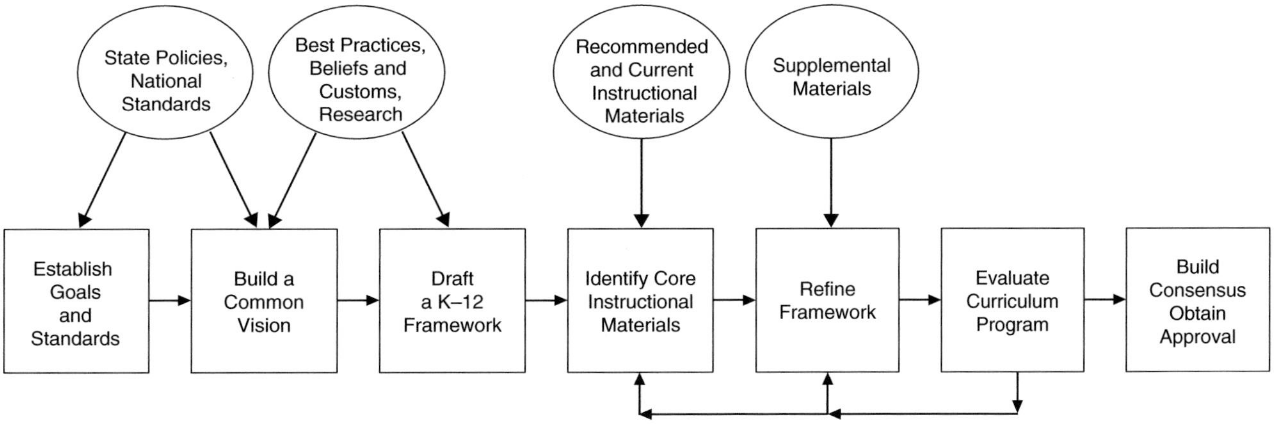

local community, community priorities, state mandates and assessments, local and state educational system structures, and local history of educational practices and programs. It is important for the committee to become familiar with this context, particularly with local, state, and national standards. Other policy documents, such as goals, mission statements, course requirements, and curriculum guides, should be considered carefully in the initial part of the design process. In addition, the committee should not only study current practices, customs, and beliefs about education in the local schools but should compare these to the educational research literature on best practices in teaching, learning, and curriculum design.

ESTABLISHING GOALS AND STANDARDS

As the starting point in the development of an improved curriculum program, a district needs goals and a set of standards to guide the work of the curriculum program design committee, particularly in the important areas of creating a framework and selecting the core instructional materials.

The previous section of the report, "Components of Coherent Mathematics and Science Education Curriculum Programs," lists criteria for goals and standards and indicates how national standards provide guidance for districts that are writing their own. In recent years, most states have adopted mathematics and science goals or standards (CCSSO, 1997). It is important for the design committee to base its work on state policy since that policy determines the extent to which state goals and standards must be used locally. Some states require local districts to follow the state standards, while others expect the standards to be used as guidelines only. In some cases, state content standards guide a state's assessment program. In these cases, districts—and their curriculum program design committees—will likely choose to focus on those standards so that their students will perform well on the state assessments.

Where local or state-level standards do not exist or where state standards are optional or do not meet the criteria for high-quality standards given in the previous section of this report, design committees may want to use national standards. Many districts and states have used the following national standards as the basis for their own standards:

- The *Curriculum and Evaluation Standards for Mathematics* (NCTM, 1989);
- The *Professional Standards for Teaching Mathematics* (NCTM, 1991c);

- The *Assessment Standards for School Mathematics* (NCTM, 1995);
- The *National Science Education Standards* (NRC, 1996b); and
- The *Benchmarks for Science Literacy* (AAAS, 1993).

BUILDING A COMMON VISION

Even with the availability of goals and a comprehensive set of standards, the curriculum program design committee needs to agree upon and articulate a common vision for the district in its own language. Teachers, administrators, and others on the committee should translate what is called for in national, state, and local standards into administrative and classroom policy and practice for their district. The committee will want to consult research literature and other sources on best practices in teaching and learning science and mathematics.

Creating a common vision of what and how students will learn mathematics and science is an important component of the development of the curriculum program, regardless of whether most of the program's components are adopted or adapted from other programs or developed independently. A common vision helps focus all stakeholders on what the school district believes is important. The vision is critical for good communication, as it will help the committee describe what the practices and behaviors of students, teachers, administrators, and parents should be when the curriculum program is in place.

In building a common vision, the design committee should describe what would be observable when the curriculum program is fully developed and implemented in terms of

- what students are learning and how they are learning it;
- what teachers are doing to support, encourage, and expect learning;
- the evidence to be used during assessment of student performance; and
- activities parents, administrators, businesses, and colleges and universities are engaged in to support and encourage high levels of student performance.

Many approaches to this part of the design committee's work are possible as long as members engage in intellectual and focused discussion regarding issues of teaching and learning. One such discussion might include tracing the development of a particular concept or strand across several grade levels, and correlating this development with national and state standards documents. Examining sample instructional materials or student work also could help the committee clarify the nature of student performances called for in the standards. Classroom examples paint pictures that communicate to teachers, administrators, supervisors, and members of the public how students engage in meaningful learning of mathematics or science. The classroom examples in the *NSES* have helped educators get a concrete sense of the national recommendations in practice. Other useful sources of examples include the NCTM's addenda to its *Standards* (NCTM, 1991a) and similar addenda in science, to be published in 1999 by the National Research Council for the Center for Science, Mathematics, and Engineering Education. In addition, committee members may want to participate in actual classroom lessons, either as teachers or as students.

It is important at this stage in the development of a curriculum program for the committee members to have an opportunity to share and discuss their many, often diverse, views about the discipline itself (mathematics or science) and about how students learn that discipline. For example, teachers may not agree about the contributions and limitations of lecture versus inquiry or of small group versus large group instruction. These issues are best addressed explicitly in a positive, professional environment where relevant educational research literature can be accessed and reviewed.

Ideally, when the vision statement is finalized, it will be a concrete statement that communicates the district's vision of what is important in mathematics and science education. It should illustrate how standards will be used and describe what a classroom should look like or what kind(s) of thinking a student should be able to do. It also should address the broader learning context, such as how teachers will teach, how students will be assessed, and how the district will support and be accountable for its students' learning.

DRAFTING A K-12 CURRICULUM FRAMEWORK

Once the goals and standards that guide the curriculum program have been identified and translated into a more specific vision of learning, teaching, assessing, and support, work on the curriculum framework can begin. This framework is meant to organize and sequence the content to create coherence in the curriculum program across all grades in all courses. It does this by assigning concepts to grade levels based on the growth and development of ideas and skills from grade to grade and by grouping the concepts to form units or courses. Whether the framework is developed locally or taken from an outside source, it should be reviewed carefully to ensure its coherence.

Although instructional materials are not part of the curriculum framework,[10] the two elements must be aligned and interconnected. Therefore, drafting a framework is the beginning of an iterative process. First, a draft of the framework indicates at which grade level concepts and skills (these may be expressed as standards or benchmarks, depending on the document being used) might be taught. Then, as instructional materials are reviewed, this early framework will be refined until a final product emerges. It may be tempting to begin assigning concepts to grade levels and/or courses immediately, but it is wise to refrain from doing so too early in the process. If districts assign concepts to grade levels or courses before looking at instructional materials, they are likely to create a need for unique units or courses. This then places them in the position of having to write materials—a difficult and expensive process that most districts do not have the resources to accomplish.

The design committee should have the following questions in mind while drafting the framework:

- Are some units so ingrained in the local curriculum that they must be retained? (If this is the case, the appropriate standards must be referenced in the framework.)
- Who are the intended audiences of the framework? How will these audiences use the framework? Will they understand it in its current form?
- What instructional resources are available locally in the community that should be built into the framework? (For example, an outdoor education laboratory school, a district greenhouse or planetarium, a museum of science and industry, or connections with a local industry may all complement a high-quality curriculum program.)

IDENTIFYING CORE INSTRUCTIONAL MATERIALS

Deciding What to Review. The goal of this step is to identify instructional materials that best support the standards as they are organized in the draft curriculum framework. As men-

[10] Because of the widespread influence of national standards in mathematics and science, increasing numbers of instructional materials and, in some cases, multi-year programs are available commercially that provide for instruction and the learning of content called for in the national standards. Both commercial publishers and curriculum development groups funded by the National Science Foundation have developed these materials. Some of the programs may address a number of the criteria for curriculum programs outlined above; therefore, local design process committees may use some aspects of these programs—as well as adjustments to their own local frameworks— to help them construct coherence within and across grades.

tioned earlier, the processes of identify-ing instructional materials and designing the curriculum framework go hand in hand. The draft framework indicates a possible assignment of standards to grade levels. Examining the instructional materials informs the committee as to what is available at each grade level, how the standards are presented in the materials, and the prior knowledge students need before beginning each unit. After the design committee reviews information about the materials, it may find a need to revise some portions of the framework.

Committee members should be selective about their review of material because they will not have time to review everything. The committee may want to become acquainted with tools developed or under development by several national groups for the identification of "exemplary" materials. For example, the National Science Foundation has made recommendations for middle-level science (NSF, 1997). The National Science Resources Center has done a similar study of materials for elementary- and middle-level science (NSRC, 1996 and 1998b). In addition, the NRC's Center for Science, Mathematics, and Engineering Education, AAAS's Project 2061, and the U. S. Department of Education all are working to analyze and identify or to help local decision makers analyze and identify exemplary mathematics or science materials (NRC, 1999c; AAAS, 1997; and DoEd, 1997).

The design committee may want to consider reviewing the materials currently in use. Even though some members will be familiar with these materials, all materials that have the potential for use in the curriculum program should be reviewed using the same set of criteria. Then, if there is a decision to drop a widely used textbook, evidence will be available. Conversely, if a district chooses to continue using a program that is unpopular, how and why this decision was made may be more easily explained.

Review Instruments and Procedures. With the introduction of content standards, most districts will need to revise their process for reviewing instructional materials, employing criteria similar to those outlined in the second section of this report. The review process should be clearly defined and involve the use of multiple techniques that give teachers and others an opportunity to analyze the materials under consideration. Supplemental materials may play a role in this process and will be discussed later in this section (beginning on pg. 37).

A considerable amount of time is needed for careful review of materials. Setting up a two-stage process has some advantages. The first stage can be used to eliminate the weakest instructional materials based on the criteria provided in this report and the design committee's vision of teaching and learning. First-stage "filtering" could involve analyzing the materials' instructional approach and quickly retaining those materials that support the skills and concepts needed for problem solving, but eliminating materials that, for example, emphasize confirmatory, pencil-and-paper activities or that would provide students with little or no experience in problem solving, data collection, and analysis. In this first stage, the committee also could exclude materials that are collections of unrelated activities or short units on single topics not connected to others. This would allow members to focus in the second stage on materials that have the most potential to support the development of a coherent curriculum program.

In the second stage, reviewers need to begin by choosing a review instrument and then practicing with it. In the practice session, all reviewers should use the instrument to evaluate the same set of instructional materials. The reviewers should discuss and agree as a group on the interpretation of each criterion to increase the reliability of the resulting reviews (i.e., independent reviewers would arrive at similar judgments). Working through this process also will give reviewers further insight into the standards that they will be using to judge the materials.

The results of the first and second stages of review should reveal strengths and weaknesses of each considered set of materials. The next step is to use this knowledge to refine the framework.

REFINING THE CURRICULUM FRAMEWORK

Finalizing Placement of Core and Supplemental Instructional Materials. The design committee, having worked with the policy documents (goals and standards), developed and described its vision of teaching and learning, analyzed the current program, and reviewed and selected instructional materials, is ready to begin refining the curriculum framework. Refining the framework involves 1) clarifying the growth of student understanding within and across years by assigning concepts to grade levels, and 2) identifying and addressing transitions and gaps in the framework, as follows:

1. Clarifying the growth of student understanding within and across years by assigning concepts to grade levels. Once instructional materials have been selected, concepts and skills can be assigned to grade levels.

When the design committee is focusing on the expected level of student understanding, they should describe performance expectations, not just specify topics. For example, a topic such as "introduction to photosynthesis" may be assigned to sixth grade. As written, this phrase can be interpreted in many ways. Some teachers may have their sixth-grade students memorize definitions; other teachers may go so far as to have students design experiments to collect evidence that plants need light; and yet others will ask students to memorize the equations that summarize the chemical reaction. In mathematics, a similar situation arises when a topic such as "length" is assigned to the first grade. Some teachers may feel that they should help students understand the concept of measuring length by using non-standard units; others may use standard units, such as centimeters or inches but not tools such as rulers and meter sticks; and yet others may go directly to measuring with rulers and meter sticks.

As discussed in the previous section on curriculum program components, the most significant difference among frameworks is their level of coherence. To build coherence into a framework requires clear descriptions of the content. Necessary prior knowledge is identified, ideas sequenced, and connections among ideas developed. This process is facilitated by using "Growth of Understanding" tables similar to Figures 4 and 5 on pgs. 26 and 27 to identify the order of introduction of concepts, both within a given year and across years.

To recap, Figures 4 and 5 illustrate how each year's instruction builds on that from prior years, with the concepts becoming progressively more complex. Such a progression helps insure that students will be able to develop a solid understanding of the "big ideas" of the discipline. Although this progression has to be considered for each major concept, skill, and ability, every concept should *not* be addressed every year. This would result in too many concepts being addressed each year without adequate time for students to develop an understanding of anything, exacerbating the problem described in TIMSS reports (Schmidt et al., 1998). Typical mathematics and science curricula attempt to address far too many topics per grade level. Unnecessary repetition can quickly stifle both student enthusiasm and understanding, leading to low expectations. One of the most important lessons from international studies and analyses cited in the introduction to this report is the value of tightening the focus of our curriculum to address fewer topics each year, allowing for much greater depth of learning.

Ideas also must build within a year. Not only should there be a lesson-to-lesson progression, but *units* should also build from one to the next. Attention must be paid to the development of themes, concepts, and skills from the beginning to the end of a school year. The framework should be explicit about how skills and concepts interweave within individual units, across a year, and over a span of years. Without this purposeful interconnection, the curriculum program cannot be coherent.

To illustrate, in science students too often read about a rigid "scientific method" at the beginning of the school year, then they read science content for the rest of the year without ever having the opportunity to experience how science concepts are derived from investigations. In mathematics, a parallel experience often occurs, when algorithmic procedures are taught as though they were important in and of themselves, rather than as tools for solving real-world problems. In such cases, the understanding that is essential to the eventual application of the algorithms is often omitted. Equally efficient procedures that are better understood—or that have been created by students—may be equally as acceptable and never considered. Even when efforts are made to address standards, concepts and skills often are separated. Students may study independent units on graphing, metric measurement, and microscope use or on plant classification, plant structure, and plant distribution, without ever seeing the connections among the concepts and skills in those units.

The following questions can guide the development of a framework that supports a logical and sequential building of student understanding:

- What prior knowledge is needed for each concept? Is it addressed in an earlier unit? Can it be presented in the current unit?
- Is a logical or developmental sequence of concepts presented within each grade level?
- How will we assist students who have missed important prior knowledge and experiences?
- How does the framework convey the importance of interconnected skills and concepts?
- How will the connections and prior knowledge and skills called for in the framework be conveyed to teachers and others who are using the curriculum program?

2. Identifying and addressing transitions and "gaps" in the framework. The instructional materials selected to support the K-12 framework may come from several sources. Ideally, one curriculum series or program would suffice for a complete span of grades, such as K-5 or K-8. The challenge is to figure out how to create links between different programs. For example, with series adopted for grades K-6, 7-8, and 9-12, the connections between grades 6 and 7 and between grades 8 and 9 must be examined closely. Does the middle-grades program pick up where the elementary grades program ends and adequately prepare students for the high-school learning goals? Linking different sources of materials makes the process of putting together a curriculum program complicated. Even when all the materials selected are of high quality, they usually differ in philosophy, program design, expectations of students, and attention to particular concepts.

Some curriculum materials are written and marketed as a set of stand-alone units that schools or districts can mix and match to meet their needs. This situation is more typical of science than mathematics, for which complete programs are available that span several grade levels. Although the mix-and-match approach gives districts more flexibility, if districts choose units from a variety of sources, care should be taken to ensure that the units achieve the desired coherence in the curriculum program.

After instructional materials have been selected, there may be places where the coherence is weak. The gaps in the curriculum program need to be identified, as well as inconsistencies in expectations of students when there are transitions from one set of materials to another. First, the big ideas (standards or benchmarks) in each program and flow between the units should be identified. If the understanding of the big ideas do not build progressively, then the remaining gaps must be filled and redundancies eliminated. It is equally important to track the skills that are expected of students, such as graphing, measurement, analysis of data, and use of tools.

A design committee refining its curriculum framework could address gaps and transition problems by recommending that

- instructional materials be modified to create transition activities;
- new units be written where major gaps exist;
- supplemental units from other sources be identified; or
- professional development be provided for teachers that addresses the issues associated with the gaps.

As mentioned earlier, modifying existing units or writing new units is an expensive and time-consuming solution to creating better transitions or covering gaps between the instructional materials from different sources. Finding units from other sources is an alternative way to fill gaps. However, this approach may result in a "patchwork" program that creates additional inconsistencies.

Professional development offers a flexible solution, in which teachers identify transition problems and consider ways to address them. They may supplement the content by using units from other programs or fill gaps or address transition problems with their own strategies.

The process described here of refining the curriculum framework may seem overwhelming. The long-term goal—to design a high-quality curriculum program for mathematics and science—is well worth the additional commitment called for. In addition, keep in mind that this report does not mention timelines because the process never ends; there will always be room for improvement. The most difficult and time-consuming part of the process is the beginning, when the design committee and the district are learning to use standards to create coherence. Once this process is understood and a few examples developed across several grade levels, continuous improvement of the total curriculum program will become much easier.

Pilot Testing the Proposed Curriculum Program. Although individual instructional materials adopted for the curriculum program may have been pilot tested in other school districts, the particular collection of materials now gathered in the design committee's framework may not have been. It needs to be. The purpose of pilot testing material is to identify problems and to correct them before full-scale implementation.

Ideally, the total program would be evaluated over several years as a cohort of students moved through the program starting in kindergarten or first grade. Obviously, full-scale implementation cannot wait for this. Rather, the district should strive to answer questions such as those listed below by piloting the program over a short period of time in several schools that together cover the total K-12 grade span.

- How well do diverse students achieve the intended results (standards) using the new instructional materials? Teacher reports and some objective test data will be useful in comparing results across schools and against desired results.
- Do students have the expected prior knowledge and skills at the beginning of each unit/course so that the coherence designed into the curriculum program can be maintained?
- When students do not have the prior knowledge and skills, how do teachers

make the needed adjustments for these students? Are any problems severe enough to justify the revision of the framework?

- What professional development is needed so that teachers can maintain the coherence of the curriculum program without denying access to students who have not attained the prior knowledge and skills called for?
- What professional development is needed to support teachers and principals in the use of the instructional materials in the curriculum program?
- What resources (time, materials, instructional technology, facilities, and community resources) do teachers need to use the materials?
- What resources are needed to maintain durability and usefulness of the materials between adoption cycles?
- What resources are needed to keep parents informed?

EVALUATING THE CURRICULUM PROGRAM

The process of developing a coherent curriculum program is not complete until the design committee has formulated a method and schedule for the periodic evaluation and improvement of the curriculum program. Like other important programmatic changes in a district, designing a curriculum program should be considered to be an ongoing process, not a one-time event. In today's rapidly changing environment, mathematics and science programs can become outdated quickly, even if they had represented state-of-the-art thinking at the time they were designed.

In making critical decisions about the nature of a new mathematics and/or science program, district policy makers and the committee will want to consider how drastic a change from their current program is desirable. As they compare their existing curriculum with national standards and their local vision, the benefits versus costs and risks will need to be weighed. An increasing number of districts also will need to take into account statewide standards of learning that are reinforced by mandatory statewide assessments. In the end, some communities may decide that they want to upgrade their mathematics and science programs dramatically to reflect the most innovative direction possible. Other communities, especially where the school district is large or where there are significant philosophical differences of opinion about what should be taught and how students should learn, may decide to proceed with more moderate change.

Dramatic change and slower, more incremental change both have their advantages and disadvantages. One advantage of dramatic change is that it requires teachers to learn new skills and ways of teaching. However, such change may be so radical that school district staff and the community reject it before there has been adequate opportunity to see results (Fullan, 1991). On the whole, a slower, more incremental approach may be more acceptable to staff and the community, but because teachers have the time and opportunity to modify new approaches, the innovation may be unacceptably compromised.

Regardless of how drastic the change decided upon it should be noted that improvement in teaching and the subsequent improvement in student achievement will take time. The implementation of an innovative program requires that teachers learn new and different teaching strategies. Often, this takes as many as three years (Fullan, 1991). Indeed, the implementation of an innovative program may result in an achievement dip during the first year or two of implementation. Furthermore, because the program advocated in this report is connected across several years, significant improvement of students' achievement likely will result from their being in the program for more than one year.

The district's evaluation plan should take into account the need for early data—as well as long-term data—on program effectiveness. The data gathered early will be helpful in deciding whether to move from pilot testing to full implementation. The data gathered over time will be critical to long-term evaluation and improvement decisions. One solution to this dual need for evaluation data is to collect enough to reach the consensus needed for formal approval of the program from the pilot sites that have been using the program for the greatest period of time and, simultaneously, to design a program evaluation process that gathers data over a more extended period of time from all schools. Such a process would provide, for example,

- a description of what constitutes a well-implemented curriculum program [11];
- a means of identifying schools and

[11] There are useful procedures for defining a well-implemented program. Three are mentioned here. First, the National Science Resources Center has developed a series of rubrics for this purpose (NSRC, 1997). Second, the Concerns-Based Adoption Model has developed a method of describing and mapping the progression of implementation of an innovative program (Hall & Loucks, 1978). Third, the National Science Foundation-funded Classroom Observation Protocols (Weiss et al., 1998; Horizon Research, 1999) provide criteria to determine effective classroom instruction and program implementation.

classrooms where curriculum programs have been well implemented; and

- a comparison of student achievement in classrooms and schools where curriculum programs have been well implemented for two or three years with student achievement in control classrooms in the same district or comparable districts. (The measures used should be aligned with the standards that were used to develop the program.)

BUILDING CONSENSUS AMONG THE STAKEHOLDERS AND OBTAINING APPROVAL

As indicated earlier, the committee responsible for designing the curriculum program will learn a great deal about curriculum, instructional materials, and other factors, such as pedagogy and assessment, that can affect the achievement of students. In the process, this committee will make a number of recommendations that will change the current program and that may impact teachers, administrators, and parents. The success of the new curriculum program ultimately will depend on the stakeholders outside the committee reaching a level of understanding and support for it comparable to that of the committee members. This requires that sustained and systematic communication be planned and executed to keep these stakeholders informed, to solicit their input, and to develop consensus and support for the committee's work. Specific strategies for the committee could include

- developing a two-way communication link with all stakeholders through status reports, newsletters, World Wide Web sites, newspaper articles, and presentations to school staffs and parent-teacher organizations;
- seeking input and response on a periodic basis through questionnaires, focus groups, use of external reviewers, and presentations to faculty and community groups;
- informing stakeholders of pilot testing results, how the feedback has been used, and the specific impact both are having on the committee's work;
- judging the degree to which the committee's work has been accepted by examining the various forms of feedback received (a committee that has communicated its work well will have a good sense of the degree to which its work will be supported); and
- convening meetings for the purpose of obtaining formal expressions of consensus.

The next step is formal approval of the curriculum program by appropriate district decision makers. When the committee has achieved consensus with other stakeholders, it is ready to request this approval. Ideally the committee would simultaneously present its implementation plan [12] and budget.

---

[12] In order to translate the curriculum program into classroom practice, a number of implementation strategies, activities, and mechanisms must be in place. These include professional development for both teachers and administrators and development of a number of support mechanisms. See Appendix A for an overview of this topic.

Appendix A

IMPLEMENTING A MATHEMATICS OR SCIENCE CURRICULUM PROGRAM

This report primarily describes the elements of a coherent curriculum program and suggests a process for its development. If the program is then to impact student learning, a number of activities must be undertaken to give life to the curriculum program through well-supported practice in all K-12 classrooms. This appendix provides a few suggestions, with a minimum of discussion, for this implementation. Particularly helpful will be the reports and studies cited here.

Support the Implementation of the Curriculum Program with Extensive Professional Development [13]

The information needed to plan full-scale implementation can be derived not only from the process used to develop the curriculum program but also from the additional processes, described in the report, of building consensus for and pilot testing the program. Typically, the implementation plan will contain the following activities [14]:

- Continued use of the information dissemination strategies of the consensus-building process noted in this report on pg. 44 to inform staff of the schedule and nature of the implementation activities and professional development they will be involved in;
- Engagement of principals and department chairs (middle and secondary level) in professional development to inform them of the nature of the new program and the results of the pilot test, as well as the following—
  - —the decisions they will need to make concerning procedures and timelines for implementation in their respective buildings,
  - —the availability and type of professional development to which their teachers will have access,
  - —the costs and sources (designers,

[13] The nature and order of the activities listed, although briefly stated, are part of an overarching strategy based on the Concerns- Based Adoption Model (CBAM) framework for understanding and responding to the concerns of teachers engaged in adopting an innovation (Hord et al., 1987).

[14] An excellent resource is *Science for All Children: A Guide to Improving Elementary Science Education in Your School District* (NSRC, 1997).

publishers) of instructional materials, kits, and equipment,
   —the support systems available to them and their teachers,
   —suggested ways of informing parents about the program and implementation,
   —the impact that the new program will have on the teacher evaluation process, and
   —plans for student and program assessment;
- Identification and purchase of the instructional materials, related kits, and equipment (Sources and methods of re-supplying expendable materials should be included in the implementation plan.); and
- Planning and execution of professional development experiences for teachers. (A comprehensive treatment of this important topic is available.)[15]

Develop and Maintain Ongoing Support and Partnerships in the Community

As important as developing consensus among district teachers and administrators is the task of developing support from and partnerships with various members of the community. These include parents of students in the school system, the non-parental public in the community, and representatives of local businesses and industry, local corporations, universities, professional organizations, science-related agencies, and museums.

Develop and Implement a Program for Assessing and Reporting Student Achievement

The wide range of reasons for examining student achievement include

- helping teachers improve their instruction and the achievement of their students;
- providing teachers with a basis for assigning grades:
- assist teachers and administrators with decisions about tracking, promotion, and graduation;
- informing students, parents, and the general public about how well students are achieving; and
- assisting in monitoring the quality of teaching, the effectiveness of the curriculum program, and other "opportunity to learn" factors (such as infrastructure support, supplies and materials availability, facilities, and so on).

The role of assessment is much broader than the assignment of grades

[15] See *Designing Professional Development for Teachers of Science and Mathematics* (Loucks-Horsley et al., 1998).

and other critical decisions about students, such as promotion and tracking. Even this traditional role for student assessment is often misunderstood and misused. (See *High Stakes Testing for Tracking, Promotion, and Graduation* [NRC, 1999a] for a careful analysis of the issues of high stakes testing that can have a major impact on students' lives.) When student assessment is considered as feedback, it is an effective means of improving student learning (Black & Wiliam, 1998) and the quality of teaching and the curriculum program. (See the *NSES* "Assessment Standards" [NRC, 1996b] for a discussion of the uses of assessment.) Of particular importance to this guide is the use of student assessment data coupled with other information in the evaluation of coherence and accessibility in the curriculum program.

Quality assessment for any of these purposes depends on the availability of effective instruments, procedures, and performance standards (see the description of performance standards on pg. 21 of this report), in addition to content standards. When performance standards do not formally exist, teachers must decide what it is that they expect students to know, understand, or be able to do. If district-wide assessment items are being developed, these become informal performance expectations.

The pervasive need for explicit performance standards that align with content standard expectations argues for inclusion of performance standards in the curriculum program. If these standards are not included in the initial design of the curriculum program, teachers and the assessment developers will need additional professional development to assist them in understanding how assessment tasks and scoring guides are developed and how performance standards are established.

Equipped with the ability to define and decide when a content standard has been met, teachers will be equipped to use assessment as a powerful means to improve their instruction and student learning.

Appendix B

MAPPING THE *NSES* FOR STRUCTURE OF MATTER (K-12)

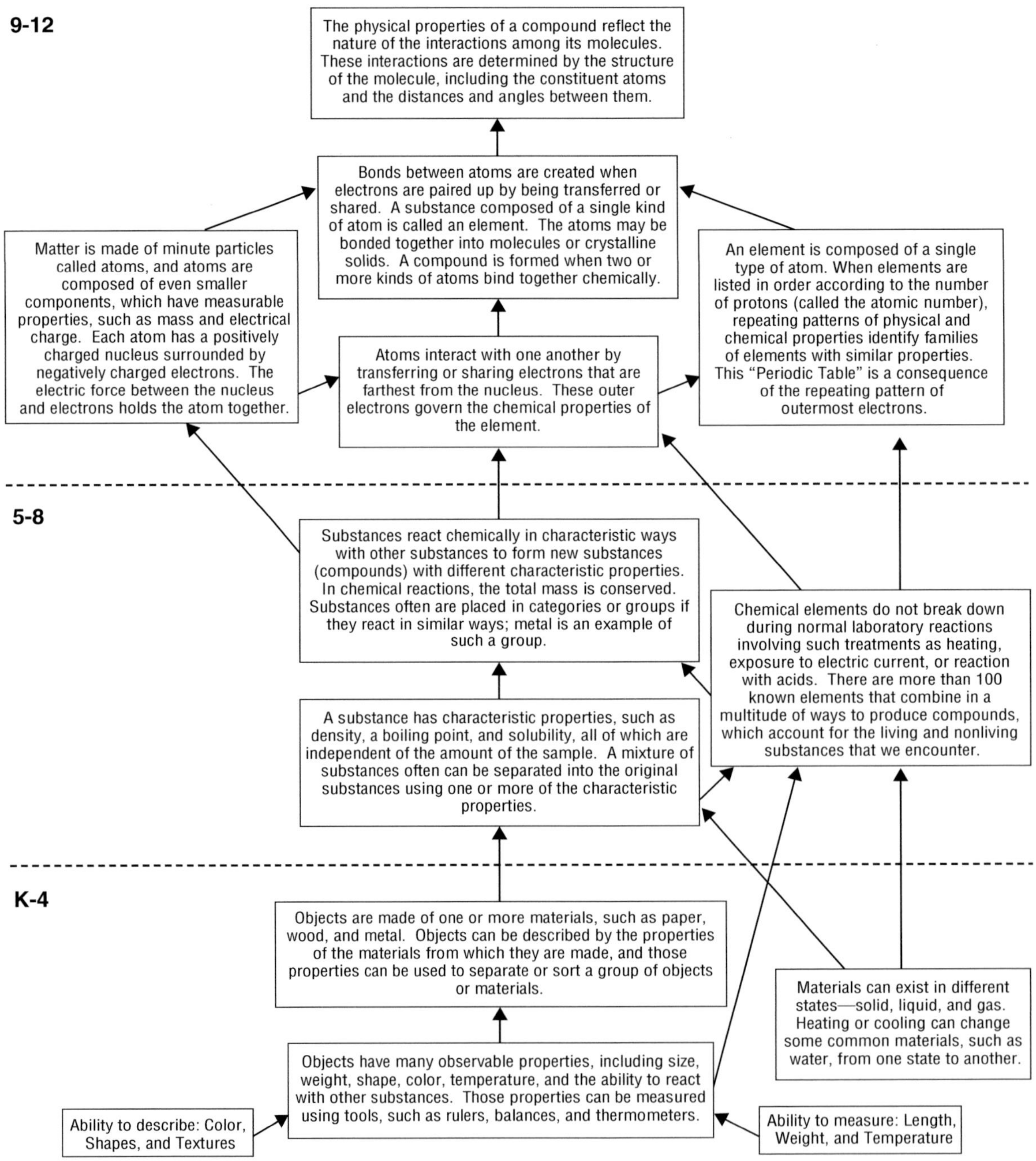

Appendix C

MAPPING THE NCTM *STANDARDS* FOR THE TOPIC OF STATISTICS (K-12)

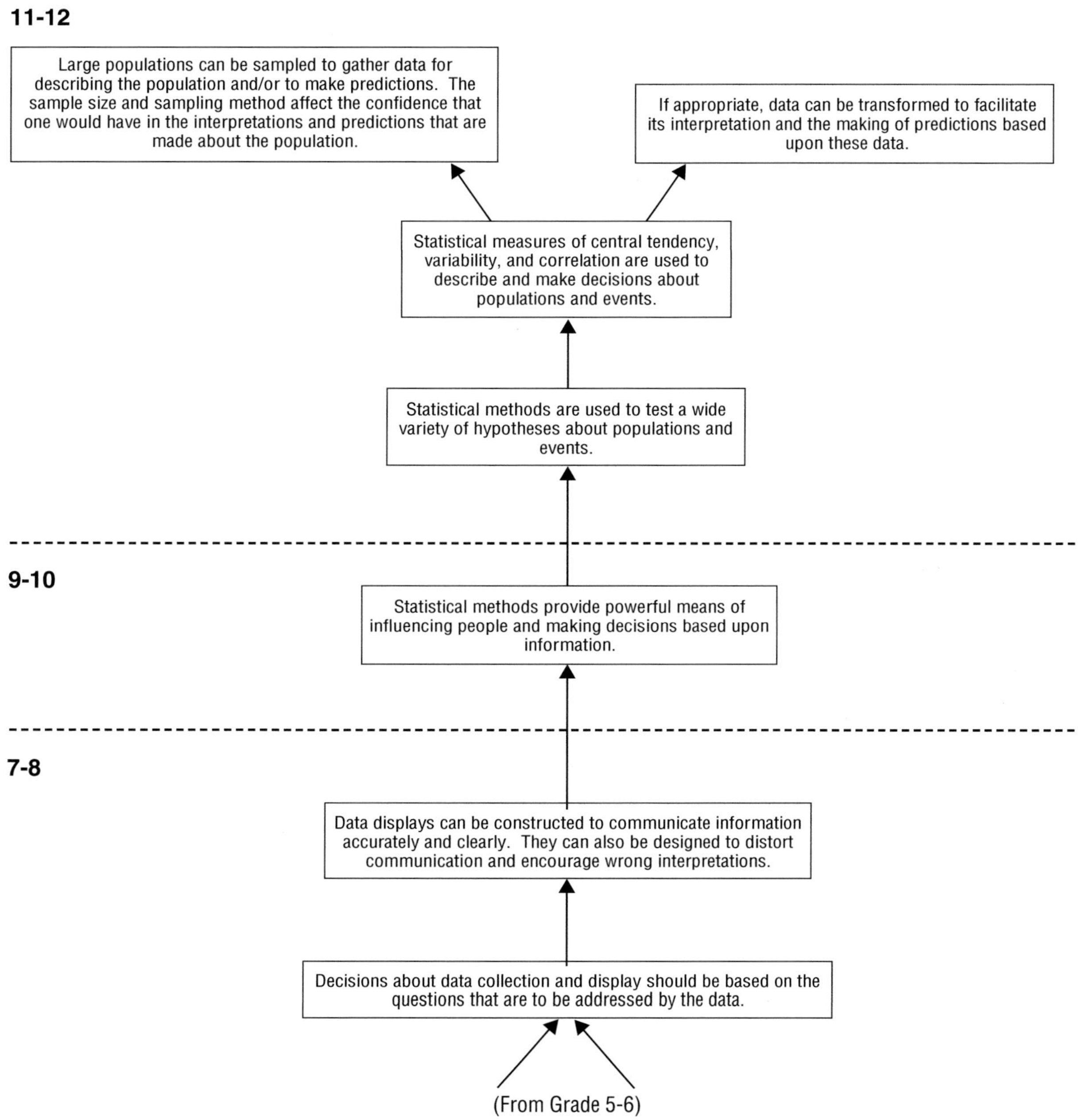

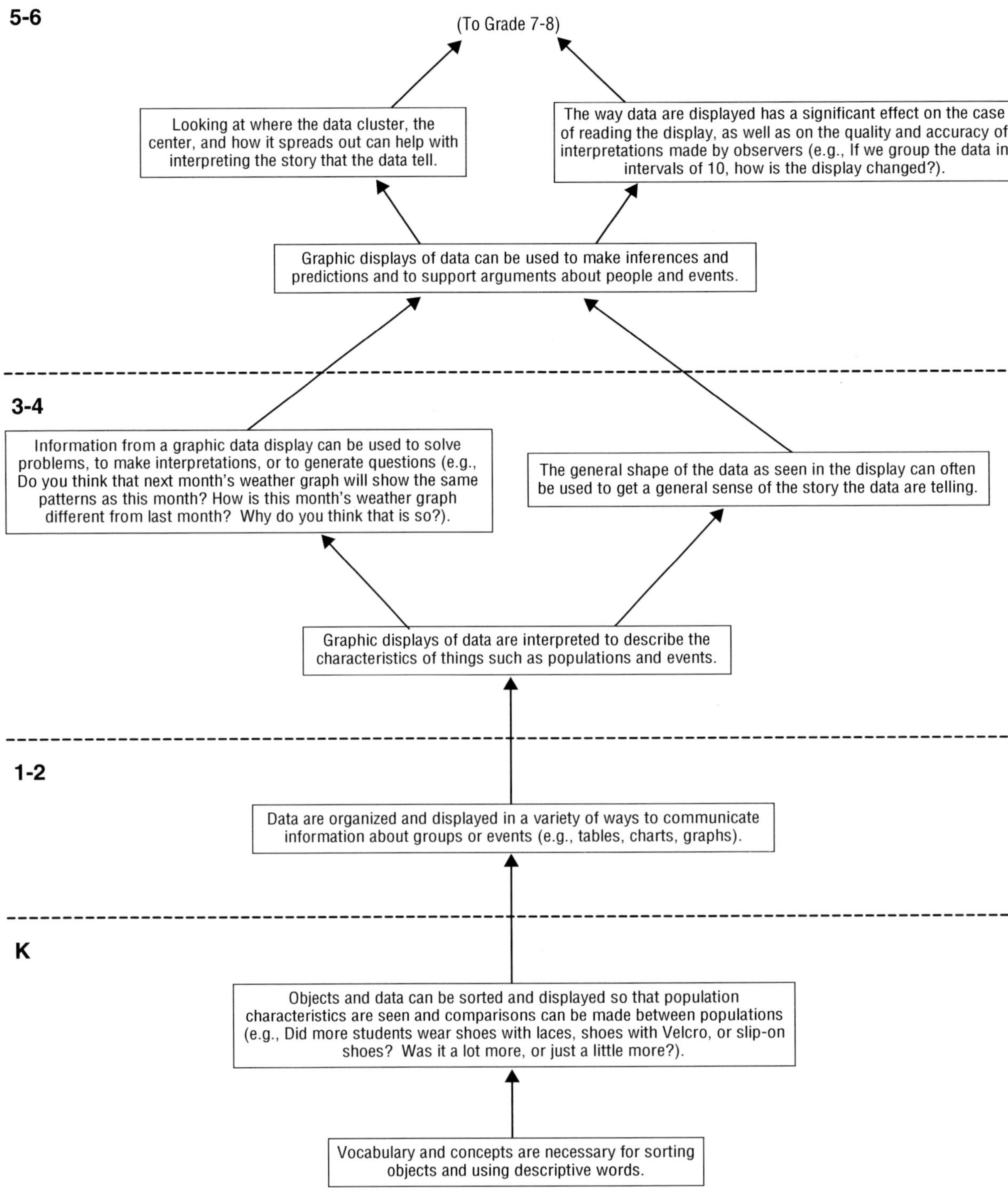
5-6
(To Grade 7-8)
Looking at where the data cluster, the center, and how it spreads out can help with interpreting the story that the data tell.
The way data are displayed has a significant effect on the case of reading the display, as well as on the quality and accuracy of interpretations made by observers (e.g., If we group the data in intervals of 10, how is the display changed?).
Graphic displays of data can be used to make inferences and predictions and to support arguments about people and events.
3-4
Information from a graphic data display can be used to solve problems, to make interpretations, or to generate questions (e.g., Do you think that next month's weather graph will show the same patterns as this month? How is this month's weather graph different from last month? Why do you think that is so?).
The general shape of the data as seen in the display can often be used to get a general sense of the story the data are telling.
Graphic displays of data are interpreted to describe the characteristics of things such as populations and events.
1-2
Data are organized and displayed in a variety of ways to communicate information about groups or events (e.g., tables, charts, graphs).
K
Objects and data can be sorted and displayed so that population characteristics are seen and comparisons can be made between populations (e.g., Did more students wear shoes with laces, shoes with Velcro, or slip-on shoes? Was it a lot more, or just a little more?).
Vocabulary and concepts are necessary for sorting objects and using descriptive words.

References

Ahlgren, A., & Kesidou, S. (1995.) "Attempting Curriculum Coherence in Project 2061." In J. A. Beane (Ed.) *Toward a Coherent Curriculum.* Alexandria, VA: Association for Supervision and Curriculum Development.

Aldridge, B. G. (1989.) *Essential Changes in Secondary Science: Scope, Sequence, and Coordination* (pp. 1, 4-5). Arlington, VA: National Science Teachers Association.

American Association for the Advancement of Science (AAAS). (At press.) *Designs for Science Literacy.* New York: Oxford University Press.

AAAS. (1999a.) *Atlas of Science Literacy.* New York: Oxford University Press.

AAAS. (1999b.) *Resources for Science Literacy: Curriculum Materials Evaluation.* New York: Oxford University Press.

AAAS. (1998.) *Blueprints for Reform: Science, Mathematics, and Technology Education.* New York: Oxford University Press.

AAAS. (1997.) *Resources for Science Literacy: Professional Development.* New York: Oxford University Press.

AAAS. (1993.) *Benchmarks for Science Literacy.* New York: Oxford University Press.

Archibald, D.A. (1998.) "The reviews of state content standards in English language arts and mathematics: A summary and review of their methods and findings and implications for future standards development." Paper commissioned by the National Education Goals Panel.

Black, P., & Wiliam, D. (1998) "Inside the Black Box; Raising Standards through Classroom Assessment" *Phi Delta Kappan*, 80(2): 139-148.

The Business Roundtable. (1996.) *Business Leader's Guide to Setting Academic Standards.* Washington, DC: Author.

Bybee, R. W. (1998.) "National Standards, Deliberation, and Design: The Dynamics of Developing Meaning in Science Curriculum." In *Problems of Meaning in Science Curriculum* (pp. 150-165). New York: Teachers College Press.

Bybee, R. W. (1997.) "Enriching the Science Curriculum." In *Achieving Scientific Literacy: from Purposes to Practices* (pp. 138-166). Portsmouth, NH: Heinemann.

Carpenter, T., Fennema, E., & Franke, M. (1992.) "Cognitively Guided Instruction: Building the Primary Mathematics Curriculum on Children's Informal Mathematical Knowledge." Madison: Wisconsin Center for Educational Research, University of Wisconsin at Madison.

Center for Technology in Learning. (1997.) "Performance Assessment Links in Science (PALS)." Menlo Park, CA: SRI International. See also http://pals.sri.com.

Connecticut Department of Education. (1998.) "The Connecticut Mathematics Program Improvement Resource Kit." Hartford, CT: Connecticut Department of Education.

Council of Chief State School Officers (CCSSO). (1997.) "Mathematics and Science Content Standards and Curriculum Frameworks." Washington, DC: Author.

Curry, B., & Temple, T. (1992.) *Using Curriculum Frameworks for Systemic Reform.* Alexandria, VA: Association for Supervision and Curriculum Development.

Driscoll, M. (1997.) "A role for mathematical story lines in the linked learning in mathematics project (LLMP)." Author's draft.

Education Commission of the States (ECS). (1996.) "Standards and Education: A Roadmap for State Policymakers." Denver, CO: Author.

ECS. (Year of publication not available.) "Mathematics and Science Standards: A Policymakers' Primer." Brochure. Denver, CO: Author.

Education Development Center (EDC). (1998.) "Choosing a Standards-based Mathematics Curriculum." Newton, MA: K-12 Mathematics Curriculum Center, EDC.

Fullan, M. (with Steigelbauer, S.). (1991.) *The New Meaning of Educational Change, 2nd ed.* New York: Teachers College Press.

Hall, G., George, A., & Rutherford, W. (1979.) *Measuring Stages of Concern about the Innovation: A Manual for Use of the Stages of Concern Questionnaire.* Austin: The University of Texas.

Hall, G., & Loucks, S. (1978.) *Innovation Configurations: Analyzing the Adoptions of Innovations.* Austin: The University of Texas.

Hoffman, K. M., & Stage, E. K. (1993.) "Science for All: Getting It Right for the 21st Century." *Educational Leadership,* 50(5): 27 -31.

Hord, S.M., Rutherford, W.L., Huling-Austin L., & Hall, G.E. (1987.) *Taking Charge of Change.* Alexandria, VA: Association for Supervision and Curriculum Development.

Horizon Research, Inc. (1999.) *Local Systemic Change: Classroom Observation Training.* Chapel Hill, NC: Author.

Gandal, M. (1997.) "Making Standards Matter: An Annual Fifty-State Progress Report on Efforts to Raise Academic Standards. "Washington, DC: American Federation of Teachers.

Glatthorn, A. A. (1994.) *Developing a Quality Curriculum.* Alexandria, VA: Association for Supervision and Curriculum Development.

Kahle, J., & Rogg, S.R. (1996.) "A Standards-based Inventory of Middle-Level Science Curriculum Materials." Interim report by authors.

Kamii, C. (1990.) "Constructivism and Beginning Arithmetic (K-2)." In T.J. Cooney (Ed.) *Teaching and Learning in the 1990's* (pp. 22-30). Reston, VA: National Council of Teachers of Mathematics.

Lappan, G., & Phillips, E.D. (1998.) "Teaching and Learning in the Connected Mathematics Project." In L. Leutzinger (Ed.) *Mathematics in the Middle.* Reston, VA: National Council of Teachers of Mathematics.

Lappan, G., Phillips, E.D., Fitzgerald, W.M., Friel, S.N., & Fey, J.T. (1998.) *Connected Mathematics.* White Plains, NY: Dale Seymour Publications (distributed by Scott Foresman Addison-Wesley).

Loucks-Horsley, S., et al. (1998.) *Designing Professional Development for Teachers of Science and Mathematics.* Thousand Oaks, CA: Corwin Press.

Loucks-Horsley, S., Kapitan, R., Carlson, M. O., Kuerbis, P. J., et al. (1990.) *Elementary School Science for the 90's.* Alexandria, VA: Association for Supervision and Curriculum Development.

Loucks-Horsley, S., & Hergert, L. F. (1985.) *An Action Guide to School Improvement.* Alexandria, VA: Association for Supervision and Curriculum Development.

Manaster, Alfred B. (1998.) "Some characteristics of eighth grade mathematics classes in the TIMSS videotape study." *The American Mathematical Monthly,* 105(9): 793-805.

McLaughlin, M. W., & Shepard, L. A. (1996.) *Improving Education through Standards-Based Reform.* National Academy of Education. Palo Alto, CA: Stanford University.

McLaughlin, M., & Shepard, L. (1995.) *Improving Education through Standards-Based Reform.* National Academy of Education. Palo Alto, CA: Stanford University.

Milbrey, W., Shepard, L. A., & O'Day, J.A. (1995.) *Improving Education through Standards-Based Reform.* National Academy of Education. Palo Alto, CA: Stanford University.

Mokros, J., Russell, S. J., & Economopoulos, K. (1995.) *Beyond Arithmetic: Changing Mathematics in the Elementary Classroom.* Palo Alto, CA: Dale Seymour Publications.

Muri, M. (At press.) "A Guide to K-12 Program Development in Mathematics." Hartford, CT: Department of Education.

National Academy of Sciences. (1998.) *Teaching about Evolution and the Nature of Science.* Washington, DC: National Academy Press.

National Association of Biology Teachers (NABT). (Year of publication not available.) "Aligning with the *National Science Education Standards*: NABT Curriculum Review Instrument." Reston, VA: Author.

National Council of Supervisors of Mathematics. (1993.) *Guide to Selecting Instructional Materials for Mathematics Education.* Reston, VA: Author.

National Council of Teachers of Mathematics (NCTM). (1998.) *Principles and Standards for School Mathematics: Discussion Draft.* Reston, VA: Author.

NCTM. (1995.) *Assessment Standards for School Mathematics.* Reston, VA: Author.

NCTM. (1991a.) "Curriculum and Evaluation Standards for School Mathematics Addenda Series." Reston, VA: Author.

NCTM. (1991b.) *A Guide for Reviewing School Mathematics Programs.* Reston, VA: Author.

NCTM. (1991c.) *Professional Standards for Teaching Mathematics.* Reston, VA: Author.

NCTM. (1989.) *Curriculum and Evaluation*

Standards for School Mathematics. Reston, VA: Author.

National Education Association. (1894.) "Report of the committee of ten on secondary school studies." Chicago: American Book Company.

National Research Council (NRC). (1999a.) *High Stakes Testing for Tracking, Promotion, and Graduation.* Washington, DC: National Academy Press.

NRC. (1999b.) *How People Learn: Brain, Mind, Experience, and School.* Washington, DC: National Academy Press.

NRC. (1999c.) *Selecting Instructional Materials: A Guide for K–12 Science.* Washington, DC: National Academy Press.

NRC. (1996a.) *Mathematics and Science Education around the World: What Can We Learn from the Survey of Mathematics and Science Opportunities and the Third International Mathematics and Science Study?* Washington, DC: National Academy Press.

NRC. (1996b.) *National Science Education Standards.* Washington, DC: National Academy Press.

NRC. (1993.) *Measuring What Counts.* Washington, DC: National Academy Press.

National Science Foundation (NSF). (1997.) "Review of Instructional Materials for Middle School Science." Arlington, VA: Author.

National Science Resources Center (NSRC). (1998a.) "NSRC Evaluation Criteria for Middle School Science Curriculum Materials." Washington, DC: Author.

NSRC. (1998b.) *Resources for Teaching Middle School Science.* Washington, DC: National Academy Press.

NSRC. (1997.) *Science for All Children: A Guide to Improving Elementary Science Education in Your School District.* Washington, DC: National Academy Press.

NSRC. (1996.) *Resources for Teaching Elementary School Science.* Washington, DC: National Academy Press.

O'Day, J., & Smith, M.S. (1993.) "Systemic Educational Reform and Educational Opportunity." In S.H. Fuhrman (Ed.) *Designing Coherent Educational Policy* (pp. 20-312). San Francisco: Jossey-Bass.

Roberts, D.A. (1982.) "Developing the concept of 'curriculum emphases' in science education." *Science Education,* 66: 243-260.

Robitaille, D. F., Schmidt, W.H., Raizen, S., McKnight, C., Britton, E., & Nicol, C. (1993.) "Curriculum Frameworks for Mathematics and Science (TIMSS Monograph No. 1)." Vancouver: Pacific Educational Press.

Roseman, J. E., Kesidou, S., & Stern, L. (1996.) "Identifying curriculum materials for science literacy: A Project 2061 evaluation tool." Commissioned paper. Washington, DC: Authors.

Schmidt, W. H., McKnight, C. C., Jakwerth, P. M., et al. (1998.) *Facing the Consequences: Using TIMSS for a Closer Look at United States Mathematics and Science Education.* Norwell, MA: Kluwer Academic Publishers.

Schmidt, W. H., & Valverde, G.A. (1998.) "Refocusing U.S. Math and Science Education." *Issues in Science and Technology.* Winter 1997-98, 14(2).

Schmidt, W. H., McKnight, C. C., & Raizen, S. (1997.) *A Splintered Vision: An Investigation of U.S. Science and Mathematics Education.* Norwell, MA: Kluwer Academic Publishers.

Schoenfeld, A.H., Smith J.P., & Arcavi, A. (1993.) "Learning." In R. Glaser (Ed.) *Advances in Instructional Psychology, Vol. 4* (pp. 55-175). Hillsdale, NJ: Lawrence Erlbaum Associates.

Smith, M.S., & O'Day, J. (1991.) "Systemic School Reform." In S.H. Fuhrman & B. Malen (Eds.) *The Politics of Curriculum and Testing.* New York: Falmer Press.

Sneider, C. I., Barber, J., & Bergman, L. (1997.) *The Architecture of Reform: GEMS and National Standards.* Berkeley: The Regents of the University of California.

Stigler, J.W., et al. (1999.) "The TIMSS Videotape Classroom Study: Methods and Findings from an Exploratory Research Project on Eighth-Grade Mathematics Instruction in Germany, Japan, and the United States." (NCES 99-074.) Washington, DC: U.S. Government Printing Office.

Stigler, J. W., & Hiebert, J. (1997.) "Understanding and improving classroom mathematics instruction: An overview of the TIMSS video study." *Phi Delta Kappa,* 79(1): 14-21.

Tyson, H. (1997.) "Overcoming structural barriers to good textbooks." Paper prepared for the National Education Goals Panel.

U.S. Department of Education (DoEd) (National Center for Education Statistics.) (1998.) *Pursuing Excellence: A Study of U.S. Twelfth-Grade Mathematics and Science Achievement in International Context.* Washington, DC: U.S. Government Printing Office.

DoEd. (1997.) *Guidelines for Submitting Mathematics Programs for Review.* Washington, DC: Author.

Van De Walle, J.A. (1998.) *Elementary and Middle School Mathematics: Teaching Developmentally.* New York: Addison Wesley Longman.

Weiss, I., Montgomery, D.L., Ridgway, C.L., & Bond, S.L. (1998.) "Local Systemic Change through Teacher Enhancement: Year Three–Cross-Site Report." Chapel Hill, NC: Horizon Research, Inc.